A History of Place in the Digital Age

A History of Place in the Digital Age explores the history and impact of Geographic Information Systems (GIS) and related digital mapping technologies in humanities research. Providing a historical and methodological discussion of place in the most important primary materials which make up the human record, including text and artefacts, the book explains how these materials frame, form and communicate location in the age of the Internet. This leads in to a discussion of how the World Wide Web distorts and skews place, amplifying some voices and reducing others.

Drawing on several connected case studies from the early modern period to the present day, the spatial writings of early modern antiquarians are explored, as are the roots of approaches to place in archaeology and philosophy. This forms the basis for a review of place online, through the complex history of the invention of the Internet, into the age of the interactive web and social media. By doing so, the book explores the key themes of spatial power and representation which these technologies frame.

A History of Place in the Digital Age will be of interest to scholars, students and practitioners in a variety of humanities disciplines with an interest in understanding how technology can help them undertake research on spatial themes. It will be of interest as primary work to historians of technology, media and communications.

Stuart Dunn is Senior Lecturer in Digital Humanities at King's College London, UK, where he has worked since 2006. He holds a PhD in Aegean Bronze Age Archaeology from the University of Durham, UK, and has interests in the history of cartography, crowdsourcing in the humanities (with a special emphasis on Volunteered Geographic Information, or VGI), and the Spatial Humanities. Most recently he has worked on the heritage of Cyprus and spatial approaches to the study of art history. His work has been funded by the AHRC, the A. G. Leventis Foundation and the Getty Foundation; and as well as being an active researcher, Dunn has developed innovative teaching modules in GIS and the humanities in KCL's MA programmes. In 2017, he co-authored the first major book-length study of academic crowdsourcing in the humanities. He has also been a Visiting Scholar in the Center for Spatial and Textual Analysis's Spatial History project at Stanford University since 2014 (USA).

Routledge Research in Digital Humanities

Series Editors: Marilyn Deegan, Lorna Hughes, Andrew Prescott and Harold Short

Digital technologies are increasingly important to arts and humanities research, expanding the horizons of research methods in all aspects of data capture, investigation, analysis, modelling, presentation and dissemination. This important series covers a wide range of disciplines with each volume focusing on a particular area, identifying the ways in which technology impacts on specific subjects. The aim is to provide an authoritative reflection of the 'state of the art' in technology-enhanced research methods. The series is critical reading for those already engaged in the digital humanities, and of wider interest to all arts and humanities scholars.

For a full list of titles in this series, please visit https://www.routledge.com/history/series/DRAH

Also in the series:

Cultural Heritage Infrastructures in Digital Humanities
Edited by Agiatis Benardou, Erik Champion, Costis Dallas and Lorna Hughes

The Strehlow Archive: Explorations in Old and New Media
Hart Cohen

Shakespeare's Language in Digital Media
Edited by Janelle Jenstad, Mark Kaethler and Jennifer Roberts-Smith

New Directions in Mobile Media and Performance
Camille Baker

The Shape of Data in the Digital Humanities: Modeling Texts and Text-based Resources
Julia Flanders and Fotis Jannidis

A History of Place in the Digital Age

Stuart Dunn

Routledge
Taylor & Francis Group
LONDON AND NEW YORK

First published 2019
by Routledge
2 Park Square, Milton Park, Abingdon, Oxon OX14 4RN

and by Routledge
52 Vanderbilt Avenue, New York, NY 10017

First issued in paperback 2020

Routledge is an imprint of the Taylor & Francis Group, an informa business

British Library Cataloguing-in-Publication Data
A catalogue record for this book is available from the British Library

Library of Congress Cataloging-in-Publication Data
A catalog record has been requested for this book

ISBN 13: 978-0-367-66150-2 (pbk)
ISBN 13: 978-1-138-22357-8 (hbk)

Typeset in Times New Roman
by Taylor & Francis Books

To my wife Emma, and the other members of my family who have learned all about academic obsession during the writing of this book.

Contents

Figures

Acknowledgments

> 'Temples (saith the auncient Aristides) are to be dedicated to the gods, and Bookes to good-men.'
>
> –William Camden, *Remaines Concerning Britaine: But especially England, and the Inhabitants thereof* (1623).

In July 2012 I had the privilege of attending a two-week NEH-funded Summer Institute in Deep Mapping at IUPUI in Indianapolis, organized by David Bodenhamer, John Corrigan and Trevor Harris. As well as introducing me to a group of like-minded spatial humanists, their collegial support and critical encouragement then and since has been invaluable to the development of my ideas about the spatial humanities. Indeed one could say that this present volume was born in the searing heat of that Indiana summer.

The Department of Digital Humanities at King's College London allowed me periods of research leave which made this book possible, and thanks must go to my various friends and colleagues there whose insights and discussions have shaped much of this book. And I also thank my students, especially the postgraduates who took my module *Maps, Apps and the GeoWeb*. We have learned together what it means to study this complicated and wide-ranging subject in Higher Education.

I would also like to thank the Center for Electronic Spatial and Textual Analysis at Stanford University, for hosting me as visiting researcher there for a period in 2014. It is a remarkable place inhabited by remarkable people.

Finally, my thanks to my editors at Routledge, Rob Langham and Heidi Lowther, for all your support, advice and encouragement, and a generosity of patience which was almost too much.

1 Spatial humanities in the digital age
The key debates

Introduction

Abstract ideas of place and space, as well as specific instances, have long formed subjects for those who write, research, create, perform and philosophize in the humanities. To understand and express *where we are* is a fundamental part of what it is to be human and, at one level or another, of all outputs of human discourse. For Aristotle, place was a container in which a distinction can be drawn between the world which contains humans generally, and that which contains the individual's body specifically, apart from the void of space and containing all things. Casey calls this latter view 'the inevitability of emplacement' (Casey, 2011). For a performer such as Michael Pearson, author of *In Comes I*, place represents an embodied experience of the landscape, to be told and retold through a variety of media (Pearson, 2015). Robert Macfarlane, who has written extensively about the experiential qualities of place, asks 'what can I know in this place that I can know nowhere else?' (Macfarlane, 2012: 27). In the history of the visual arts, individual practitioners became closely associated with specific landscapes, such as Van Gogh with the landscape of Arles, and Turner with Kent. Writers too appropriate particular lands and cityscapes, such as Thomas Hardy's imagining of Wessex, or Arthur Conan Doyle's Victorian London. The significance of spatial representation is undiminished in imagined worlds, bridging real life and literary landscapes. As a political writer, George Orwell invokes place as a metaphor, or rather a microcosm, for political aspiration, the 'fruitful fields of England' of *Animal Farm*, where a very specifically English rural idyll contrasted with allegorical reference to the ideologies of Communism for dramatic and political effect. Similarly, for historians and archaeologists, place frames and provides essential referents for interpretation of the past, with many of the key archaeological and historical debates of the mid and late twentieth century focusing on whether past societies shaped, or were shaped by, the places around them.

Place, therefore, matters greatly in all branches of the humanities; and humanists have responded to this in a variety of ways. Among the most important recent engagements between the humanistic domains and ideas around place are development and deployment of digital mapping and spatial analysis technologies. From the 1990s onwards, the latter have drawn critical and methodological attention across the humanities to the importance of place as a product of human thought, and as a tool for intellectual enquiry. This collection of methods is a key basis of the *spatial humanities*. This umbrella term refers to the methods and processes which facilitate the critical investigation of place, space and location as an artefact of human history and experience. While, I would argue, the spatial humanities do not necessary have to involve computational methods, these have certainly helped to define and channel many of the most important philosophical and analytical enquires around place in the humanities since at least the turn of the present century. By the same token, the spatial humanities differ from cognate fields such as human geography and spatial analysis, as the humanist's fundamental concern is the primary assets on which the traditional humanities are built, such as texts, manuscripts, old maps, material culture, and performative environments.

These are the media which convey human discourse and cognition from human to human, and from the past to the present.

The idea that place has a special importance for such human discourse is not new; indeed it is over forty years since Yi-Fu Tuan described 'humanistic space' as those places in our lives which are occupied (and defined) by the parameters and limits of perception and experience, rather than by artificial boundaries and delimitations (Tuan, 1976). For Tuan, place is what people made it through living and experiencing, regardless of scale or any external notion of significance. In his famous phrase, '[a]n arm-chair by the fire is a place, but so is the nation-state' (Tuan, 1976: 269). However, when one writes about place as an author, creates an representation of a place as an artist, encodes spatial information about objects (such as archaeological find spots) as metadata, or engages with place through social media, that spatial or place 'object' becomes detached from its creator's experience, and transmitted to many further locations via channels of broadcast, publication and other media, to points of interaction with readers and writers across space and time.

The key premise of this book is that the spatial humanities are the observation of this process of transmission, and that to update the idea of 'humanistic place' for the twenty-first century we must understand both how digital technologies enable the transmission and analysis of spatial information; and also how they fit into much longer histories of human place-making. Analysis using the functionalities of software such as Geographical Information Systems (GIS) is a relatively new phenomenon, but it makes no sense to assert that the types of 'humanistic place' they engage with began with the invention of those technologies. This book seeks to trace the history of these ideas, and the place of technologies such as GIS within that history.

A new label for an old problem

The term 'spatial humanities' first appeared as a subject-area label in 2010, with the publication of *The Spatial Humanities: The Future of GIS in*

Humanities Scholarship, a volume of ten chapters edited by David Bodenhamer, a social historian with strong connections to the social sciences, the religious scholar John Corrigan, and Trevor Harris, a professor of archaeology with strong interests in GIS (Bodenhamer, Corrigan, & Harris, 2010). The interdisciplinary nature of these scholars' expertise, and the methodological concern with place which links them, reflects the newly labelled field's profoundly interdisciplinary nature. Spatial research can draw on qualitative and quantitative social science data, and humanistic data, raising critical questions as to what the concepts of 'qualitative', 'quantitative', 'data' and so on mean in those domains. Since 2010, driven in no small part by such questions, the spatial humanities have become increasingly visible, especially in Europe, North America, and to Asia, marked by further published collections including *Toward Spatial Humanities: Historical GIS and Spatial History* (Gregory & Geddes, 2014) and *Literary Mapping for the Digital Age* (Cooper, Donaldson, & Murrieta-Flores, 2016). It has also seen the establishment of centres of research and teaching dedicated to the spatial humanities including (but certainly not limited to) Stanford University's Spatial History project, the University of Kent, Lancaster University's spatial humanities hub, and IUPUI's Virtual Center for Spatial Humanities.[1] It features as a series of the Indiana University Press,[2] and in major Digital Humanities degree programmes such as the MA in Digital Humanities at King's College London, to give but one example.[3]

The spatial humanities, as manifested as a branch of the literature, discourses and institutional environments of the humanities more broadly, have been accelerated and, to an extent, defined, by *technological* developments – most notably the emergence of GIS as a tool used by humanists to explore their research questions. As also noted, there has been a marked shift from application-oriented analysis of particular problems using GIS (and related technologies), to broader and more unified analyses based on general theories of spatiality. For example, once applications making use of GIS to interrogate and analyse historical information were termed 'Historical GIS', which implicitly emphasizes the *technology* of GIS with case studies tacked on and the 'History' acting as a qualifier – the area where it happens to have been applied – these now fit more easily under the more methodologically

1 Stanford Spatial History – http://web.stanford.edu/group/spatialhistory/cgi-bin/site/index.php; Kent's Interdisciplinary Centre for Spatial Studies (KISS) – https://www.kent.ac.uk/sac/research/research-centres/kiss-secondary.html; Digital Humanities Centre, Lancaster – http://wp.lancs.ac.uk/dighum; Polis Center at IUPUS http://polis.iupui.edu/index.php/spatial-humanities/project-1; Spatial Humanities press series in Indiana – http://www.iupress.indiana.edu/index.php?cPath=1037_3130_3700; KCL's Digital Humanities degree – https://www.kcl.ac.uk/study/postgraduate/taught-courses/digital-humanities-ma.aspx

2 http://www.iupress.indiana.edu/index.php?cPath=1037_3130_3700

3 https://www.kcl.ac.uk/study/postgraduate/taught-courses/digital-humanities-ma.aspx

integrated term of 'Spatial History' (Gregory & Geddes, 2014). In the same way, 'Literary GIS' as a device for the exploration of spatial references within fiction may now fit under the broad term of 'literary cartography' (Rossetto, 2013). This is what Kelley Anne Knowles, in a review of Gregory and Geddes, 2014, termed a 'shift from empirically focused HGIS to the more general concept of spatial history' (https://southernspaces.org/2014/has-historical-gis-arrived-review-toward-spatial-humanities). This represents a methodological shift towards the study of general principles and broad understanding, as opposed to answering particular research questions. However, despite these moves in individual domains within the humanities, there remains a lack of a cross-domain, and cross-discourse analysis that brings together conceptions of place which come to the present-day reader/viewer/audience/user via all the major media and content of the humanities. More importantly in the present context, there remains a relative lack of investigation in to where, when and how these ideas' deeper history emerged.

Despite this, the emergence of the spatial humanities on to the institutional academic scene in these ways reflects its growing depth as a subject of critical enquiry in its own right, often driven by GIS and related technologies. In particular, as noted, Gregory and Geddes (2014) stress that the spatial humanities represents a shift away from the application of GIS tools to particular research problems and questions, towards a coherent intellectual framework through which those technologies can be critically applied. In the terminology of Harold Short and Willard McCarty, when they were framing Digital Humanities as a discrete field, this view of spatial humanities is more akin to a 'methodological commons' (McCarty, 2003). There remains, however, no single satisfactory definition for what the spatial humanities are, still less a concrete or shared understanding of what such a field's epistemology might look like.

Most humanities disciplines are defined, whether explicitly or not, by their relationships with their materials of study, rather than by methodology: texts for historians and literature scholars, objects for archaeologists, performance for theatre studies scholars, and so on. So follow the trajectories described above, where scholars in all these fields are becoming increasingly focused on 'place', is 'place' really a suitable *subject* of study in its own right? 'Place' despite a long history of theorization (Cresswell, 2015), does not have an ontological cohesion or physical presence of its own, in the way that other materials studied by humanists (such as text) do. Individual examples of applications which might be uncontroversially described as examples of spatial humanities research can be drawn from the fields of text (in its various forms of prose, poetry, drama etc.), (historical) cartography (Southall & Pridal, 2012), the visual arts (Daniels, 2004), physical artefacts and museums (Moser, 2010; Moser, 2007), landscape studies and archaeological sites (Wheatley, 2004), and performance (Pearson, 2015). In all these, places can be both a subject of discourse and a product of it.

Such applications mark what has been called the *spatial turn* in the humanities (Warf & Arias, 2008). Following other such 'turns' (the 'visual turn', the 'cultural turn' etc.) marking points at which certain concepts took on their own significance as areas of study in particular sections of the literature, the emergence of the spatial turn may be traced to the 1990s, with a *floruit* from the mid-2000s to the present day (Goodchild & Janelle, 2010). It was in the period leading up to this, coinciding with the launch of web mapping platforms like OpenStreetMap (OSM) in 2004 and Google Earth in 2005, that publications presenting applications of GIS technologies in the humanities began to appear. It was also the time when the limitations of such technology for understanding place in the human past and present started featuring in debate.

The spatial humanities and the limits of shallow mapping

The very term 'mapping' is a fraught and complex one, with overtones of delineation, demarcation, power, imbalance and subjectivity. Mapping as conventionally understood in terms of representing horizontal, linear features on the Earth's surface, represents that blend of subject and product which, for the purposes of this book (and following Tuan) we will call 'humanistic place'; but for many of the kinds of humanistic applications referred to above, it is politically, culturally and intellectually unsatisfactory. This is especially so if one takes a definition of mapping which stresses spatial quantification as a means of reduction for the purposes of navigation, planning and wayfinding; of reducing the complexity of the landscape. For example, Andrews (2009: 1) defines a map as '[a] partly schematized graphic representation whose signs are interrelated in ways that resemble, or could reasonably be thought to resemble, the horizontal relations connecting the objects represented.' He goes on to argue that '[s]pace relations can be described by words like "adjacent to" or "north of", but a sentence formed from these and other verbal utterances could never count as a map, because it does not sufficiently resemble the object it refers to' (ibid.).

This definition may be correct, but the spatial humanities, defined as the critical treatment of place in the various types of discourse identified above, necessarily requires the distinction between a map as Andrews describes, and 'space relations' to be blurred. Some scholars have identified a set of methods, theories and practices aimed at such a blurring as 'deep mapping'. Deep mapping has similarly been the subject of a good deal of literature recently, much of it focused on the limitations of the map as described by Andrews for discursive, qualitative material. It is a logical antonym to 'shallow mapping', which is equated with the kind of superficial schematization in cartography apparent in such definitions. A 'deep map' has many possible attributes, but is defined by Bodenhamer et al (2015: 3) as: '[A] finely detailed, multimedia depiction of a place and the people, animals, and objects that exist within it and are thus inseparable from the contours and rhythms of everyday life … [s]imultaneously a platform, a process and a product' and '[a] way to engage evidence within its spatiotemporal context and trace pathways of discovery, framed as a conversation and not a statement'.

A recent issue of the journal *Humanities* dedicated to the subject defined deep mapping in similarly broad terms:

> The elemental and very reasonably put question 'what is deep mapping?' is best tackled not by outlining a set of defining characteristics and features (an exercise that is unavoidably weighted by the ballast of disciplinary persuasion) but by surveying the various precincts by which, as a coagulation of approaches and (inter)disciplinary interventions, it is performatively put to work. (Roberts, 2016: 2)

The reference to 'ballast of disciplinary persuasion' directs us to acknowledge the great diversity of different kinds of place which exist in the humanities, as outlined above, but similarly to be aware that any satisfactory definition of deep mapping must be methodologically robust enough to survive transmogrification across the various branches of the humanities. This means avoiding preoccupation with the disciplinary output, or product, and focusing on the process. A key point for both definitions is that a deep map is something that is *done*, not just something that is designed, published, authored, read and consumed in the manner of a traditional schematic map. For despite its limitations, mapping of this type is historically the primary vehicle for communicating place. The spatial humanities, in particular its methodological incarnation of deep mapping, invite us to question schematic mapping as a means of communicating humanistic place; and spatial humanists are thus beginning to explore alternatives. This will prove key to the ideas around spatial narrative that are developed later in this book.

The concept of 'depth' in the present discourse on 'deep mapping' is problematic. A deep map brings representational *desiderata* together in multiple ways, some made visually, some implicit, some hidden altogether. Articulation of these basic elements leads to an acknowledgement that the production of a schematic map as a visual object is part of a process of representing a space, of which only part is the portion of the Earth's surface it occupies. For example, a traditional, schematic map may be produced by the UK's government mapping agency, the Ordnance Survey, of a particular area. This will be produced according to strict methodologies, using fixed scales and representational conventions for things like buildings, forests, mountains and watercourses (Oliver, 2005). It has no beginning, middle or end in a narrative sense. However, in a deep mapping, or spatially critical sense, it acquires these properties *when it is used by an individual reader*. If they use it to plan a journey, that journey has a start, a route, way-markers, and a finish. If the reader travels this route, then it becomes embodied, and enriched with their experiential processes, with spatial (as well as literal) footprints on and around the route. Texts attesting to places within those footprints could be added. And so on. When one is dealing not with an Ordnance Survey map, but with an ancient object or manuscript where the methods of capture and recording of place are less explicit and less methodologically clear, then further spatial-critical scaffoldings are needed.

Other parts of the process include survey, selection of what to represent and what leave out, the gathering and assimilation of non-visual types of data such as place names and (for certain kinds of map) textual descriptions, and of course, the process of reading the (deep) map itself. It is a short conceptual step from such a process of knowledge construction, to that long familiar to students of narrative structure and literary criticism. Notions of authorship, expression and narrative construction also acknowledge that the cognition of place and the construction of place cannot be separated: As with Doreen Massey's observation in Kilburn (see Chapter 2), and Kwan and Ding's analysis of the trajectories of female Muslim populations after the September 11th attacks in the US in 2001 (Kwan & Ding, 2008), cognition and construction of place are made at a particular point in space and time, both of which can be described and categorized – and quantified and qualified – in 'scientific' terms. In the spatial humanities, an equivalent process must be undertaken by the contemporary digital cartographer who might be removed from the point of platial cognition and construction by many centuries, or even millennia. It is here that certain intellectual processes of the humanities – such as the approaches of rhetoric and phenomenology – can be called upon.

Bodenhamer et. al. (2010) are undoubtedly correct to assert that a 'deep map' is process rather than a product. The examples cited here of applications of Qualitative GIS and Critical GIS in human geography are based on methodologically robust processes of quantification and qualification. The spatial humanities require cognate processes to handle the complexities of mapping socially constructed place from historical and archaeological source material. For example, what would a map look like if expressed as a bibliography (the main aim of the Cyprus corpus described in Chapter 7), and vice versa? When an author describes a location or a space on the Earth's surface, in prose or verse, how can that description be rendered into a spatial footprint which is displayable on a map in positivist terms? And how do the scripting languages that underlie the digital representation of geographic data enable the combination of text and maps in a way that enables new and/or different readings of geographic space? These are all questions of critical engagement and narrative building.

This is not to deny the value for the humanities of maps that are not 'deep'. Harris explicitly rejects the term 'shallow map' as an antonym for that of 'deep map': 'intentionally or otherwise [shallow map] implies a meaning of superficiality and inconsequentiality and that they are lightweight and lacking in substance. There is overwhelming evidence to disprove these … descriptions' (Harris, 2015). This description of the superficiality of a 'shallow map' echoes the reductive criticisms of GIS in its earlier applications to the humanities, especially archaeology (Lock, 2001, see also Leszczynski, 2009a). The implication is that a spatial narrative, or a deep map, must involve a close reading of text, context, image and cartography at the same time, augmented by non-quantitative understandings and processes.

From spatial turn to spatial literacy

At the same time, the flood of digital spatial information now available in the world makes critically understanding and receiving that information ever more difficult. The set of skills required for doing so has been termed *spatial literacy* (Bodenhamer & Gregory, 2011), a term which has found some penetration into mainstream humanities classroom activities, such as the teaching of Classical Studies (e.g. Schindler, 2016). The need for spatial literacy in the modern world extends beyond the humanities and academia, but the emphasis on the need to dig deeper than the 'shallow' or linear realms of spatial quantification is markedly similar. Policymakers and major global organizations such as the World Bank have recognized a discursive and critically informed understanding of place and space as crucially important to good decision making (Sui & DeLyser, 2012: 113). In 2008, the government of the UK published a landmark document, *Place Matters: The Location Strategy for the United Kingdom*, which noted that 'Policy-makers have realized the crucial importance of space and place in understanding the complexity of the world's problems, seeking solutions to these problems that will work well under diverse local circumstances'. The report notes that the UK's national mapping agency, the Ordnance Survey, directly or indirectly underpinned over £100bn of GDP (UK Geographic Information Panel, 2008: 8, 14). This reflected the way in which critical consideration of space surfaced in the academic humanities in the same period.

The importance of spatial literacy teaching and learning at all levels has been recognized in the university classrooms of the humanities. Core contributions, such as Bodenhamer and Gregory (2011), place spatial literacy at the heart of spatial humanities pedagogy (Bodenhamer & Gregory, 2011). The ability to understand space, and how technology, including the increasingly mobile internet deals with and shapes it, is important for a range of life scenarios in an age of digital culture, where most citizens are, at some point, regularly guided in their daily actions by Internet-delivered spatial media and information. As Bednarz and Kemp (2011) state, 'spatial literacy is becoming tightly linked with citizenship'. Linda L. Hill describes three kinds of geographic knowledge, which transpose easily to spatial literacy: *Declarative*, which is the acquisition and retention of empirical spatial facts (e.g. London is part of England); *Procedural*, which refers to the ability to navigate between multiple points (e.g. how to walk to King's College London from Charing Cross station); and *Configurational*, which is the ability to infer complex arguments or statements from spatial information (e.g. working out the optimum way of getting from one point to another accounting for an individual's requirements, such as the need to divert to a shop to buy a particular type of a particular product) (Hill, 2009: 22–3). In parallel with the discussion above (and following) on deep mapping, these place an emphasis on *doing* and *acting* in place, rather than simply observing it,

Spatial literacy has also been defined in terms of the ability of humans to navigate geography beyond its quantitative limits. Goodchild defines this as

'[The ability of an individual to] capture and communicate knowledge in the form of a map … understand and recognize the world as viewed from above' (Goodchild, 2006). But he goes on to argue that spatial literacy is also the ability 'to recognize and interpret patterns, know that geography is more than just a list of places on the Earth's surface, see the value of geography as a basis for organizing and discovering information, and comprehend such basic concepts as scale and spatial resolution' (ibid.). This emphasis on the capturing and reception of place reflects the view, advanced above, that the spatial humanities are about the critical observation and understanding of place in human discourse. A better, critical understanding of a world which, as will be seen, is heavily regulated by spatial information, thus depends very muchon a better and humanized understanding of spatial data itself. This is entirely compatible with the vision of spatial humanities set out in this book.

Digital humanities; spatial humanities

Returning to the shift from application- (and project)-oriented research to epistemologically grounded theory, the spatial humanities reflects the last decade's history of the broader Digital Humanities (DH). DH has also undergone a shift from being a techno-centric and instrumentalist support-oriented activity termed 'Humanities Computing', to one of the detached critical enquiry associated with an academic field (Svensson 2009); in a way which mirrors the shift form 'Historical GIS' to 'Spatial History'. Much of the reason for this lies in methodological crossover or 'methodological commons' (McCarty, 2003); whereby the application of computational methodologies crosses domains, yet whose application differs according to those domains' critical requirements. John Unsworth, in his 2000 overview of the field of Digital Humanities, expresses these methods in terms of seven 'scholarly primitives', activities which form the basic building blocks of scholarly work: discovering, annotating, comparing, referring, sampling, illustrating and representing.[4] Like the Digital Humanities, the spatial humanities are concerned with all of these, yet for the spatial humanities, a *cross-domain* critical framework is not yet established.

Take for example the application of the Text Encoding Initiative, the XML standard that forms the staple of many Digital Humanities textual research activities (Romanello, Berti, Boschetti, Babeu, & Crane, 2009). Defined simply as 'guidelines for encoding machine-readable texts in the humanities and social sciences',[5] TEI facilitates the encoding of individual elements of text according to what they are (e.g. places, persons, objects, etc.), rather than how the text is presented on a webpage. It has a set of cross-domain applications, which leads to the questioning of text as a building block of humanities discourse. TEI has been applied to mark-up text in fields as diverse as Early

4 http://www.people.virginia.edu/~jmu2m/Kings.5-00/primitives.html
5 http://www.tei-c.org/About/faq.xml#body.1_div.1_div.1

English Laws, the works of Samuel Johnson, and of ancient Greco-Roman inscriptions. Digital humanists, however, tend to focus the application of TEI to a body of text on the *process*, rather than the *product*. In Porsdam's words 'TEI resulted in an interest in the representation of information and knowledge coming from the humanities as a topic in its own right' (Porsdam, 2011). The same is also true if one considers recent projects involving the crowdsourced application of TEI markup to bodies of text, such as UCL's Transcribe Bentham Project, where it was noted that a user-friendly TEI markup tool 'would relieve volunteers from being overly-concerned with encoding and allow them to concentrate upon deciphering Bentham's manuscripts, and result in the submission of a greater number of transcripts at a faster pace' (Causer, Tonra, & Wallace, 2012: 133). This emphasis on process as well as production invites parallels with the concept of the deep map described above, with critical approaches to spatial literacy, and with the more discursive and critical models of Digital Humanities that have recently emerged (Berry, 2011). This also highlights the fact that Digital Humanities research as opposed to humanities computing is, like deep mapping, something that is *done* and understood, not simply *produced* – a process, not a product, much like deep mapping.

The subtle difference between communicating place and representing it is one of the key areas of 'methodological commons' which underpins the spatial humanities; and we can think of the latter process – representation – as 'the digitization of place' (Dunn, 2017). The complexity of digitization is well understood across the Digital Humanities as a process which involves the assessment of assets, the processes of reading and interpretation, and the construction of a nuanced digital artefact; characteristics which, arguably, much spatial information on the WWW currently lacks.

A Digital Humanities project setting out to digitize a text would be broken down into several stages. It would begin with digital capture through scanning, followed by a process of converting the text into machine-readable form, whether through an automated process of Optical Character Recognition (OCR), manual transcription, or whether the task is devolved to a wider group through crowdsourcing (Causer et al., 2012), and error correction. This process would likely include markup into XML/TEI, again through an automated or manual process which would, in either case, require close expert supervision and review. As in the case of epigraphic digitization projects such as the *Inscriptions of Roman Tripolitania*, there would be a process of annotation and editing, and the addition of *apparati critci*. Textual digitization is the production of digital artefacts through a closely nuanced interplay of quantitative and qualitative processes, where the sequence of letters (for example) may (or may not) be quantitatively correct in that they correspond directly between the printed page and the screen; but there are also layers of qualitative and interpretative material. This lies somewhere between the first and second 'waves' of the Digital Humanities, as described by Presner in 2010:

> [T]he first wave of Digital Humanities scholarship in the late 1990s and early 2000s tended to focus on large-scale digitization projects and the establishment of technological infrastructure … [T]he current second wave of Digital Humanities – what can be called 'Digital Humanities 2.0' – is deeply generative, creating the environments and tools for producing, curating, and interacting with knowledge that is 'born digital' and lives in various digital contexts. Presner, quoted in (Berry, 2011).

The present problem for the spatial humanities is how humanistic place, conceived as a distinct set anthropogenic categories, yet following the logic of Tuan, intersects with Berry's 'third wave' of Digital Humanities: 'the underlying *computationality* of the forms held within a computational medium' (Berry, 2011: 4, emphasis in original). The digitization of place, whether this is place contained in humanities discourse (see above), the 'humanistic place' of Tuan, or the experiential place of the GPS-enabled smartphone user, or data shared on platforms such as Google Maps, is rendered connectable through digitization, but it is not 'deeply generative'. How it becomes so is what distinguishes the 'spatial humanities' from various discipline-oriented prefixes of the term 'GIS' which has characterized the field in the past. It is also an intellectual world which has been profoundly impacted by developments in digital mapping technologies outside the relative methodological security of GIS applications.

The GeoWeb: the Internet's own spatial turn

By far the most important of these is, of course, the development of digital mapping on the World Wide Web (WWW). Most importantly for the present discussion, it has blurred the distinction between *communicating* (information about) place and *representing* it. The period in which both deep mapping and the spatial humanities emerged – the 2000s onwards – is also the period in which place and the WWW came together. A large body of literature on the history of the internet in this period has stressed that the creation and distribution of spatial data online mirrored the process of the Internet's development, most notably a shift in emphasis from *form* to *content*. The WWW of the 1990s was based on HTML which determined the size, font, emphasis and layout of content. However, with the development of XML, and the consequent ability to represent the content of data and not just its form, the emphasis shifted to describing and representing data content according to its structure. It will be argued in detail in the next chapter that the spatial humanities can only be understood in the context of the history of the Internet and the WWW, in the context of types of spatial communication that went before.

Place is represented on WWW platforms in very specific ways (of which TEI is but one familiar to digital humanists: see above, and Chapter 4); but the most basic unit of place on the WWW is the decimal latitude/longitude number string, referencing to the World Geodetic System (WGS). The WGS, an ellipsoid general referencing system for world geography, was certified as a general standard by the US Defense Mapping Agency in 1984, based on the global geometry of weighted least squares (Decker, 1986), providing a means of referencing georeferenced data systematically across a variety of local geodesies; and which made it amenable for deployment with 'digital products', including the Internet, whose key breakthroughs coincided with this period of technological history.

The Internet itself began in 1969 as an experiment in collapsing space – allowing computer users to undertake the previously impossible task of running programs on geographically distant machines that were coterminously (Abbate, 2000). Further experimentation in the 1970s led to standardization of the software required for this process of remote operation (Cerf & Kahn, 1974), making it seamless and essentially masking the fact that the user was accessing from an entirely differently location. Place and space were thus collapsed, and merged both experientially and conceptually. The present-day GeoWeb – a concept explored in more detail in Chapter 2 – is essentially a subset of this worldwide linked infrastructure, which is distinguished by its particular representation of precisely (but not necessarily accurately) located and horizontally related features on the Earth's surface.

The present GeoWeb is based on Linked Open Data (LOD) protocols, which emerged (and continue to emerge) by consensus from community governance organizations such as the W3C, the Internet Society and the Open Geospatial Consortium (OGC) (Lake & Farley, 2009). The GeoWeb merits its special designation among the broad, connected worlds of LOD by virtue of the special characteristics of spatial data, and the shared concept of place it represents. This is reflected in the well-defined sub-class of Semantic Web ontologies that describe place (see, for example, the GeoNames web ontology at http://www.geonames.org/ontology/documentation.html; and the review of place ontologies in Ballatore, 2016). Given the oft-quoted (and possibly apocryphal) statistic that 80 per cent of all content on the WWW is georeferenced in some way, the Web can be said to have undergone a 'spatial turn' of its own, at the core of which is spatial Linked Open Data.

This spatial turn of the WWW is underscored by the relatively recent emergence of Global Positioning Systems (GPS) as a dominant technology in mobile communications. Once a tool available only to militaries and governments, GPS is now a ubiquitous feature of mobile devices. Acquiring location information using GPS-enabled smartphones is now the most common task for which American customers use their devices (Anderson, 2016). General usage of the WWW *in place*, while making use of, and contributing to, the spectrum of online geospatial data represents a major shift in the user experience of the WWW as a whole. Applying the critical frameworks of the spatial humanities and deep mapping to this mass of geodata may not only, as described above, assist in the fostering of spatial literacy and good citizenship, but also provide valuable new research questions for the spatial humanities themselves. For example, the way in which GPS data connects with information about heritage sites opens up new ways in which smartphones can be used to read space and create individualized narratives (Dunn & Schumacher, 2016); a phenomenon explored in more detail in Chapters 6 and 8. Some researchers argue that the mobile web affords some of the most exciting future possibilities for fields such as Literary GIS and reader-generated mapping (see Cooper & Priestnall, 2011). However, like all applications of LOD, the power of GPS on the GeoWeb lies in its ability to connect, search, synthesize, rank and, to an extent, visualize (Berners-Lee, Hendler, & Lassila, 2001); for example by the sharing of GPS traces on the OpenStreetMap platform (see https://www.openstreetmap.org/traces). It does not enable close quantitative or qualitative spatial analysis, and thus lacks the essential investigative qualities of a GIS. Redressing this balance will surely be one of the key challenges for the spatial humanities over the next ten years.

The humanities and GIS

Another key challenge highlighted by the growth of the GeoWeb is the dichotomy between spatial narrative and spatial analysis. The analytical platform provided by GIS, and the capacity to perform quantitative analyses across raster and vector datasets in humanities contexts, is the antecedent of the spatial humanities. As noted above, in the (recent) past 'GIS' has acted as a suffix for various branches of the humanities concerned with the computational analysis and investigation of space and place – Historical GIS, Literary GIS, Archaeological GIS etc. Accordingly, it is one of those classes of technology with which the spatial turn in the humanities is most closely associated, but in many ways has struggled to throw off the limitations of technological determinism. This has prompted humanities scholars engaged with GIS to probe those limitations. In the literature that has explored the spatial turn, GIS is no longer a specific set of technologies to enable spatial analysis of precisely quantified map data (of which more below); today it is taken as an umbrella term to describe a whole suite of software, standards, methods, tools, applications, standards and approaches which allow the analysis and visualization of spatial relationships. As the website of ESRI, one of the principal proprietary suppliers of GIS products puts it, GIS allows us to 'visualize, question, analyze, and interpret data to understand relationships, patterns, and trends' (see http://www.esri.com/what-is-gis). This is why humanists have co-opted it as part of a generative framework for analysing their data, and why, as Bodenhamer et al. have stated, 'GIS lies at the heart of this so-called spatial turn' (Bodenhamer et al., 2010: vii).

The attractions of GIS for spatial humanists are practical as well as epistemological. Widespread availability, lowering costs and technical barriers to uptake, as well as the emergence of free Open Source GIS products such as QuantumGIS (Hugentobler, 2008) have without question been responsible for the upsurge in the interest in place as a subject of study in the humanities in the last ten years. The attendant emergence of more specific sub-terms such as

Humanities GIS, Literary GIS, Archaeological GIS, Historical GIS etc., serve only to highlight the importance of the connected suite of technologies and approaches that GIS represents in the humanities more broadly; as does the process, noted above, of their assimilation into the general discourse of Digital Humanities within those fields.

GIS and GeoWeb: different beasts?

Because of the relative fluidity of the term 'GIS' in scholarship, and the fact that it can no longer be taken to refer purely to a type of software, it is important to make certain distinctions as to what GIS is *not*, or at least how the meaning of the term has changed in the past ten years. Analytical GIS must be differentiated from the GeoWeb as defined above, which refers to data or services which have geographical referents (such as a latitude and longitude) being published in a linkable way on a network, via a platform such as Google Earth (see above). This basic truth notwithstanding, the GeoWeb and GIS share a reliance on information mediated into data that is plottable (itself a loaded term) into *Cartesian space*, which implies absolute fixity according to an extrinsic geometrical framework. Cartesian space, deriving from the writings of René Descartes (1596–1650, Latin *Cartesius*), refers to information plotted on to a horizontal plane, with parameters defined by two axes. Cartesian space is inherently representational and positivist. In the 1990s both the humanities and the critical-theoretic branches of geography shared a reaction against the (supposed) positivism and reductionism of GIS analytical approaches, a rejection of the positivistic precepts of 'spatial science', and an assertion of post-processual and post-structuralist critical paradigms.

It is worth at this point turning to some of the approaches adopted by human geographers to the problem of quantifying Cartesian space. Leszczynski (2009b: 357) seeks to problematize the quantitative/qualitative distinction by describing the distinction between discursive human-interpretative views of the world and the computation data structures upon which GIS rests as a divide between the 'infological' and 'datalogical'. These two views of the world, Leszczynski argues, cannot necessarily be resolved, but they can interact producitvely. The main risk of applying a Cartesian approach to spatial humanities research questions lies in the assumptions inherent in converting discursive platial entities into computational form using some form of geo-digitization. Consider the Oxford English Dictionary definitions of two key terms: system (as in Geographic Information System), and programme (as in computer program): 'An organized or connected group of things' in the ninth sense of the former case, 'A sequence of operations that a machine can be set to perform automatically' (both from www.oed.com). Both definitions stress the linear, sequential and 'datalogical' nature of constructing knowledge, whereas understanding that process, and the reasonings behind each step, is the 'infological' aspect – the process of understanding place and space from the discourses of the past, as opposed to the act of producing a map, or other reductive analysis.

Like human geography, the spatial humanities deals with the geographical exploration of both quantitative and qualitative information. As Leszczynski argues, place and space may be divided into quantitative and qualitative categories: absolute space or place which refers to a particular 'footprint' on the Earth's surface which can be mapped and measured, and relative space or place which is referred to abstractly, such as a place name (Jiang & Yao, 2006; Hill, 2009). If we accept that the spatial humanities is a result of the use of *both* GIS and GeoWeb technologies to explore humanistic space, then this distinction between the ways in which place is represented – and transformed into digital data – becomes critical, in both senses of the term. Indeed, we may adopt the nuanced and context-aware vision of GIS adopted by Thatcher et al., 'Critical GIS' (Thatcher et al., 2015), a key epistemology for the spatial humanities.

This brings us back to what, in the humanities, is 'mappable' in the terminology of Andrews (see above), and what is 'unmappable'. Schuurman has argued that 'cultural space' and 'knowledge maps' employ a complex and non-linear transformation to physical space, and that the two are ultimately 'unmappable' in conventional terms (Shuurman, 2005). This goes to the heart of the tension between human space, which is experienced, felt, lived and moved through, and abstract space, which can be measured, computed, and subjected to quantitative analysis. Critical and Qualitative GIS are responses to the need to map the latter, feelings, experiences, perceptions – human cultural spaces in the present world – make them mappable by employing rigorous qualitative frameworks. The humanities, on the other hand, are fundamentally concerned with 'cultural space' and 'knowledge maps' which emerge from primary sources from which the observer/geographer/humanist might be separated my many centuries, requiring a different set of interpretative steps.

On the other hand, the Critical and Qualitative branches of GIS have long been recognized as 'a set of practices and social processes' which, as such, are open to contestable analysis, and to appropriation by other traditions such as Participatory GIS and feminist geography (O'Sullivan, 2006). It remains true that the technical affordances of what it means to represent place in digital form have been relatively little explored in the humanities, at least in a manner where methods and approaches can be appropriated by other humanities domains (a gap in the literature made more apparent by the extensive work done in the parallel worlds of Critical GIS, Critical Cartography and Digital Geography: see O'Sullivan, 2006; Sheppard, 2005; Pickles, 2006). Rather than the present distinction between 'quantitative' and 'qualitative' data, which have a concrete basis in human geography but less so in the spatial humanities (see above, with reference to the discussion about spatial humanities and Digital Humanities), it is more useful to think of how we use digital methods to investigate place from the past, compared with place in the present. Much of the spatial humanities – which emerges from the methodological commons formed by spatial history, literary cartography etc. – needs a critical framework for the former, but in the process might also enrich the latter through better dialogue with the established approaches and practices of Critical and Qualitative GIS in the social sciences.

Conclusion: from spatial humanities to deep mapping

In their discussion of spatial thinking in archaeology, Lock and Pouncett state that '[a]s a process of reasoning [deep mapping] draws into focus the tension between models and narrative, GIS models as the essence of academic endeavour to uncover meaning waiting to be discovered and narrative as the emergent complexities within story telling' (Lock & Pouncett, 2017: 6). They distinguish between 'spatial analysis', as the kinds of formal methods and models of GIS, and 'spatial narrative', as the less formal way of spatial story telling associated with Web 2.0 – the spatialized aspects, which I described above under the heading of the GeoWeb, an entity that is absolutely essential for framing the spatial humanities, and yet which emerged through a complex blend of historical accident and designs which had nothing at all to do with the field or its methodologies.

Throughout this discussion however, it is assumed that both GIS and the GeoWeb place the same kind of epistemic and ontological pressures on spatial representation. To exist in either, the places/spaces in question must be reduced to a digital point or set of points. This shared ontological genesis of GIS and the GeoWeb is critical to bear in mind when both are considered as sets of technological affordances which support humanistic research and human spatial literacy – and place the stress on the processes of what we do with them, rather than the outputs. This goes to the heart of why the 'spatial turn' in the humanities is driven by technology. The spatial humanities are defined by the basic aim of generating primary knowledge about space and place in the material and intellectual records of human history. In doing so, they are framed by three factors, all inextricably linked with the affordances of the digital medium: *representation, communication* and *analysis*. These themes will be returned to in the conclusion.

Interrogating the past using both spatial analysis and spatial narrative (to stick with Lock and Pouncett's distinction) while recognizing that both rely on the fundamentally reductive process of representing place as digital media (be that GIS shapefiles, latitude/longitudes/KML file and so on) involves moving beyond representational maps to post-representational cartography. This is not merely a problem for GIS in archaeology. Rossetto, for example, has called for a 're-cartographization' of the field of literary mapping, where mapping text moves beyond the simple representation of places mentioned or attested therein, to a more exploratory use of spatial technique to understand those references. She notes that '[t]he idea that maps come to life when they are practised in particular settings and situations is crucial for a post-representational approach to cartography' (Rossetto, 2013: 10).

A key contention of this book is that the spatial humanities should represent the study of spatial analysis and spatial narrative in a way which, first and foremost, separates the creator of humanistic space from its consumers, and then problematizes that separation. It is also essential to recognize that this process is not unique to digital technology of either the narrative or the analytical stripe, rather humans have been approaching the description of place in this way for centuries. In doing so, it builds on recent syntheses in human geography, which have identified a shift towards 'hybrid geographies', a '(re)turn to synthesis and holism … Hybridizing, remixing, and mashing up conceptual frameworks, data sources, and modes of analysis as these works do may provide a means to cross the methodological, epistemological, and philosophical chasms that have divided human geography' (Sui & DeLyser, 2012). However, 'holism' is only possible (and useful) if the different parts which make up the holistic whole are properly understood.

In the following chapters, it is argued that (a) GIS and (b) the visual and ontological vocabularies of data representation on the GeoWeb can be leveraged as critical methods for spatial humanists to interrogate the human record, past and present, more effectively. These methods are intrinsically linked with how the contemporary world understands and perceives space: they have grown with the Internet and the WWW, a process of rapid technological change over the last forty years or so, which has led to what Massey has called 'space-time compression', the 'shrinking' of the world by communications technology, conditioned by access and wealth (Massey, 1991). There can be no understanding of spatial narrative of the past using digital methods without an appreciation of the way in which GeoWeb technologies have moved (and can move further) beyond positivism and reductionism, and where they fit in to a much deeper and longer history of spatial research and expression. This historical context is explored in Chapter 2, beginning with the cartographers and antiquarian chorographers of the sixteenth century onwards. This historical perspective is further explored in Chapter 3, where the instance of archaeology is considered as a shorter case study. Archaeology's roots in spatial thinking go particularly deep, and with Chapter 2, this provides an overview of what one might think of as the *longue durée* of the spatial turn in the humanities – a prehistory of spatial thinking which in fact goes back many centuries before the internet and WWW focused scholarly (and wider) attention on place in the way it has today.

Text, however, is a prime concern of most humanities domains, and it must have a central place in any discussion of the spatial humanities. The point-based gazetteer is only one approach to quantifying place in text, and this has drawn little on the theoretical advances of Literary GIS, and has made fairly limited forays beyond the idea of the gazetteer entry as a point (on a map) to, for example, the gazetteer entry as an event (Mostern & Johnson, 2008). Chapter 4 seeks to develop links between these ideas, and those of the Digital Humanities which, as noted above, are characterized by a progression from linear to discursive approaches to digital textual scholarship. For the textual spatial humanities, a model of 'biblio-textual' representation of place is suggested, which draws on existing models of citation, referencing, authority and abstraction to break place down into 'mappable' components.

Chapter 5 seeks to link the discursive narratives about the spatial humanities which form the historical discussions of Chapters 2 and 3, and the textual ones of chapter 4, with current debates on user-generated spatial data, or 'neogeography'. It explores questions about collective mapping in the form of Volunteered Geographic Information, or 'volunteered geography' on platforms such as OpenStreetMap. It is shown how the ideas behind the GeoWeb contribute to the notion that spatial narrative is a means of enactment, not simply something that is subjectively related or presented. The ontological structure of place on the GeoWeb reflects that of the 'mappable' components of humanities discourse described above. This provides a key methodological intersection between the spatial humanities and digital debates within human geography – in fact, it is arguable here that we see the idea of 'the human' and 'the digital', set out in Dunn (2017) acting as an epistemological link between the two traditions. It is concluded here that the contemporary GeoWeb engendered unique types of partnership between academia and the public, which can potentially enable new forms of spatial research.

Chapter 6 tackles directly the idea of spatial narrative. Building from the premise set out here that the digital environment provides unique forms of communicative place, this chapter attempts a high level typology of the ways place is communicated in humanities research.

However it is constructed, spatial narrative is a transmedial means of expressing spatial significance, which is based on existing models of citation, referencing, gazetteers, authority and abstraction to break place down into structures. This is the topic of Chapter 7. For example, many projects in digital Ancient World geography, which deals with the text-based Classical literature canon, build gazetteers from places referenced in those texts by annotating them, and building tools to explore those annotations (Simon, Barker, Isaksen, & de Soto Canamares, 2015). The structuring of spatial information, which links the GeoWeb, traditional library approaches, and contemporary debates about power and colonialism, are explored, with special reference to a recent project which uses digital mapping to represent the heritage of Cyprus – a profoundly contested area of humanistic and geographical space – as expressed in maps and text.

Many of the arguments in Chapters 1–7 deal with the problems of representing physical engagement, especially the discussions of phenomenology in archaeology in Chapters 3 and 6. Chapter 8 addresses more directly the idea of the spatial humanities and physical movement and embodiment, using two case studies to explore how digital methods can be used to describe and explicate dynamic human movement at both landscape and micro scale. This addresses distinctions that have been made in previous literature between spatial narrative and other narrative forms such as film, books etc., which often stress the fact that the latter do not arise from physical engagement and movement, whereas the former does. Chapter 8 examines this assertion by assessing critical approaches to mobility and dynamism, and how digital methods can help explain these in both the past and the present. The chapter explores how movement narratives can be captured, analysed and interrogated and – in the interdisciplinary spirit framed above – highlights ways in which visual artists have played with this tension.

The concluding chapter returns to the idea of deep mapping as a family of methods, some digital, some not, which can offer new insights in to a range of areas in the humanities by communicating, analysing and connecting places. As noted above, deep maps have been presented as a solution to the problem that these spaces are complex, fuzzy, inconsistent, and often do not lend themselves well to 'traditional' Cartesian representation – which, as is clear from what has gone before, is much of the spatial information dealt with by humanists. However, simply to consider this as a dichotomy between 'deep' and 'shallow' mapping is to deny the complexities of the representation of humanistic place, as conceptualized by Yi-Fu Tuan (see Harris, 2015). A more useful question to ask is what critical frameworks are required for deep, processual, post-representational ways of exploring place in human discourse, from the past to the present. Just as with robust and replicable GIS models which archaeologists have used to explore data about the past (see Lock, 2010; Lock & Pouncett, 2017), and just as there are critical frameworks for encoding text in TEI (Romanello et al., 2009); deep mapping requires a critical framework for the embodied representation of geographic data encoded within the humanities. The various elements of this critical framework are explored in this volume.

I conclude this chapter by returning to the key question for all researchers of humanistic place employing digital methods – practitioners of the spatial humanities – whether ancient, historical or modern: 'what does it mean to *digitize a place?*' While spatial humanities and human geography share a concern with both the epistemology of place (the means and methods of its investigation) and its ontology (how it is constructed and presented) (Cresswell 2015), and how the distinction between the two might be deconstructed or even collapsed altogether (Leszczynski 2009b), both begin from a critical appreciation of what it means to 'map' as a verb, something that is done, rather than just *is.* It is important therefore to appreciate that 'digitizing [a] place' – or expressing place in digital form – and digital mapping are related, but different, things.

References

Abbate, J. (2000). *Inventing the Internet.* MIT Press.

Anderson, M. (2016). More Americans using smartphones for getting directions, streaming TV. Pew Research Center. Available online: http://www.pewresearch.org/fact-tank/2016/01/29/us-smartphone-use/

Andrews, J. H. (2009). *Maps in those days: cartographic methods before 1850.* Four Courts Press.

Ballatore, A . (2016). Prolegomena for an ontology of place. In H. Onsrud & W. Khun (eds.), *Advancing Geographic Information Science: the past and next twenty years* (pp. 91–103). GSDI Association Press.

Bednarz, S. W. & Kemp, K. (2011). Understanding and nurturing spatial literacy. *Procedia – Social and Behavioral Sciences*, 21, 18–23.

Berners-Lee, T., Hendler, J., & Lassila, O. (2001). The semantic web. *Scientific American. Sci Am*, a division of Nature America, Inc.

Berry, D. M. (2011). The computatonal turn: Thinking about the digital humanities. *Culture Machine*, 12,1–22.

Bodenhamer, D. J., Corrigan, J., & Harris, T. M. (2010). *The spatial humanities: GIS and the future of humanities scholarship.* Indiana University Press.

Bodenhamer, D. J., Corrigan, J., & Harris, T. M. (2015)*Deep maps and spatial narratives.* Indiana University Press.

Bodenhamer, D. J. & Gregory, I. N. (2011). Teaching spatial literacy and spatial technologies in the digital humanities. *Teaching Geographic Information Science and Technology in Higher Education*, 231–246.

Casey, E. (2011). Do places have edges? A geophilosophical inquiry. *Envisioning Landscapes, Making Worlds: Geography and the Humanities*, 65–73.

Causer, T., Tonra, J., & Wallace, V. (2012). Transcription maximized; expense minimized? Crowdsourcing and editing The Collected Works of Jeremy Bentham. *Literary and Linguistic Computing*, 27(2), 119–137.

Cerf, V. G. & Kahn, R. E. (1974). A protocol for Packet Network Intercommunication. *IEEE Trans on Comms*, 22(5). Available online: http://eolo.cps.unizar.es/docencia/doctorado/Articulos/ArticulosFamosos/1974 - Vinton Cerf - ProtocolForPacketNetworkIntercommunication-IEEE TComms.pdf

Coffee, K. (2007). Audience research and the museum experience as social practice. *Museum Management and Curatorship*, 22(4), 377–389.

Cooper, D., Donaldson, C., & Murrieta-Flores, P. (eds.). (2016). *Literary mapping in the digital age.* Abingdon: Routledge.

Cooper, D. & Priestnall, G. (2011). The processual intertextuality of literary cartographies: critical and digital practices. *The Cartographic Journal*, 48(4), 250–262.

Cresswell, T. (2015). *Place: an introduction*. John Wiley & Sons.

Daniels, S . (2008). Landscape and art., in J. Duncan, N. C. Johnson, & R. H. Schein (eds.) *A companion to cultural geography*, pp. 430–46. John Wiley & Sons

Decker, B. L. (1986). World Geodetic System 1984. Available online: http://www.dtic.mil/docs/citations/ADA167570

Dunn, S. (2017). Praxes of "The Human" and "The Digital": Spatial humanities and the digitization of place. *GeoHumanities*, 3(1), 88–107.

Dunn, S. & Schumacher, M. (2016). Explaining events to computers: Critical quantification, multiplicity and narratives in cultural heritage. *Digital Humanities Quarterly*, 10(3).

Goodchild, M. (2006). The Fourth R? Rethinking GIS Education. |*ArcNews*, 28(3), 1, 11.

Goodchild, M. F., & Janelle, D. G. (2010). Toward critical spatial thinking in the social sciences and humanities. *GeoJournal*, 75(1), 3–13.

Gregory, I. N. & Geddes, A. (2014). *Toward spatial humanities: historical GIS and spatial history.* Indiana University Press.

Harris, T. M . (2015). Deep geography – deep mapping. Spatial storytelling and a sense of place. *Deep Maps and Spatial Narratives*, 28–53.

Heffernan, J. A. W. (1991). Ekphrasis and representation. *New Literary History*, 22(2), 297–316.

Hill, L. L . (2009). *Georeferencing: The geographic associations of information.* MIT Press.

Hugentobler, M. (2008). Quantum GIS. In *Encyclopedia of GIS* (pp. 935–939). Springer.

Jiang, B. & Yao, X. (2006). Location-based services and GIS in perspective. *Computers, Environment and Urban Systems*, 30(6), 712–725.

Kwan, M.-P. & Ding, G. (2008). Geo-narrative: extending geographic information systems for narrative analysis in qualitative and mixed-method research. *The Professional Geographer*, 60(4), 443–465.

Lake, R. & Farley, J. (2009). Infrastructure for the geospatial web. In *The Geospatial Web* (pp.15–26). Springer.

Leszczynski, A. (2009a). Poststructuralism and GIS: Is there a 'disconnect'? *Environment and Planning D: Society and Space*, 27(4), 581–602.

Leszczynski, A. (2009b). Quantitative limits to qualitative rnanagements: GIS, its critics, and the philosophical divide. *The Professional Geographer*, 61(3), 350–365.

Lock, G. (2001). Theorising the practice or practicing the theory: Archaeology and GIS. *Archaeologia Polona*, 39, 153–164.

Lock, G. (2010). Representations of space and place in the humanities. In D. Bodenhamer, J. Corrigan, & T. M. Harris (eds.), *The spatial humanities: GIS and the future of humanities scholarship* (pp. 89–109).

Lock, G. & Pouncett, J. (2017). Spatial thinking in archaeology: Is GIS the answer? *Journal of Archaeological Science*, *84*, 129–35.

Macfarlane, R. (2012). *The old ways: a journey on foot*. London: Penguin Books.

Massey, D. (1991). A global sense of place. *Marxism Today*, 35,315–323.

McCarty, W. (2003). Humanities computing. *Encyclopedia of Library and Information Science, 2*, 1224.

Moser, S. (2010). The devil is in the detail: Museum displays and the creation of knowledge. *Museum Anthropology*, 33(1), 22–32.

Mostern, R. & Johnson, I. (2008). From named place to naming event: Creating gazetteers for history. *International Journal of Geographical Information Science*, 22(10), 1091–1108.

Oliver, R. R. (2005). *Ordnance Survey maps: a concise guide for historians.* Charles Close Society for the Study of Ordnance Survey Maps.

O'Sullivan, D. (2006)Geographical information science: critical GIS. *Progress in Human Geography*, 30(6), 783–791.

Pearson, M . (2015). *In comes I: Performance, memory and landscape.* Royal College of General Practitioners.

Pickles, J. (2006)Ground truth 1995–2005. *Transactions in GIS*, 10(5) 763–772.

Porsdam, H. (2011). Too much 'digital', too little 'humanities'? An attempt to explain why many humanities scholars are reluctant converts to Digital Humanities. Available online: https://www.repository.cam.ac.uk/handle/1810/244642

Roberts, L. (2016). Deep mapping and spatial anthropology. *Humanities*, 5(1), 5.

Romanello, M., Berti, M., Boschetti, F., Babeu, A., & Crane, G. (2009). Rethinking critical editions of fragmentary texts by ontologies. *The 13th International Conference on Electronic Publishing*, 155–174.

Rossetto, T. (2014). Theorizing maps with literature. *Progress in Human Geography*, 38(4), 513–530.

Schindler, R. K. (2016). Teaching spatial literacy in the classical studies curriculum. *Digital Humanities Quarterly*, 10(2). Available online: http://www.digitalhumanities.org/dhq/vol/10/2/000252/000252.html

Schuurman, N. (2005)Social perspectives on semantic interoperability: Constraints on geographical knowledge from a data perspective. *Cartographica: the International Journal for Geographic Information and Geovisualization*, 40(4), 47–61.

Sheppard, E. (2005)Knowledge production through critical GIS: genealogy and prospects. *Cartographica: The International Journal for Geographic Information and Geovisualization*, 40(4), 5–21.

Simon, R., Barker, E., Isaksen, L., & de Soto Canamares, P. (2015). Linking early geospatial documents, one place at a time: Annotation of geographic documents with Recogito. *E-Perimetron*, 10(2), 49–59.

Southall, H. & Pridal, P. (2012). Old maps online: Enabling global access to historical mapping. *E-Perimetron*, 7(2), 73–81.

Sui, D. & DeLyser, D. (2012). Crossing the qualitative-quantitative chasm I: Hybrid geographies, the spatial turn, and volunteered geographic information (VGI). *Progress in Human Geography*, 36(1), 111–124.

Svensson, P. (2009)Humanities computing as digital humanities. *Digital Humanities Quarterly*3(3). Available online: http://www.digitalhumanities.org.dhq/vol/3/3/000065/00065.html

Thatcher, J., Bergmann, L., Ricker, B., Rose-Redwood, R., O'Sullivan, D., Barnes, T. J., Barnsemoore, L. R., Beltz Imaoka, L., Burns, R., Cinnamon, J., Dalton, C. M., Dunn, S., Harvey, F. and Jin-Kyu, J. (2015). Revisiting critical GIS. *Environment and Planning A, Economy and Space*, 48(5), 815–824.

Tuan, Yi-Fu (1976)Humanistic geography. *Annals of the Association of American Geographers* 66(2) (June), 266–276. Published by Taylor & Francis Ltd. on behalf of the Association of American Geographers. Available online: http//www.jstor.org/stable/2562469.

UK Geographic Information Panel. (2008). *Place matters: the Location Strategy for the United Kingdom. Communities.* Retrieved from https://data.gov.uk/sites/default/files/uk-location-strategy_10.pdf

Warf, B. & Arias, S. (2008). *The spatial turn: Interdisciplinary perspectives.* Abingdon: Routledge.

Wheatley, D. (2004). Making space for an archaeology of place. *Internet Archaeology*, 15.

2 The *longue durée* of the spatial humanities

Part I: Communicating place

Introduction

As noted in Chapter 1, knowing where we are in space and time is a fundamental human requirement. Different forms of human discourses create, present and communicate place in different ways. In this context, the information revolution precipitated by the Internet and, later, the World Wide Web (WWW) ushered in a new suite of narrative and analytical approaches to the communication of place and led (as argued by Bodenhamer et al. (2010)) to the spatial turn in the humanities and elsewhere. However, this process did not *begin* with the Internet or WWW. Digital technologies have merely enhanced and expanded the processes of spatial communication, as other new forms of media did before them.

Like many other forms of place-making that went before and since, the digital construction and communication of place is a function of communications media and of the limitations of those media, and how the processes of engaging with them – such as reading or viewing – relate to the local, regional and global scales of abstract geography. In his short story, *On Exactitude in Science,* Jorge Luis Borges tells the story of an empire which fantastically negates the tensions of scale by imagining an empire which maps itself at a scale of 1:1, where the '[m]ap of the Empire whose size was that of the Empire, and which coincided point for point with it'. The debates stimulated by the advent of GIS, and related web mapping technologies in the humanities, are merely the latest of those in hundreds, if not thousands, of years of humanity grappling to represent place in the media available to it.

For human views of place and space have always been inextricably linked to human technical capacity, especially the capacity to *document* and *communicate.* Spatial thinking, spatial analysis and spatial literacy are therefore inextricably intertwined with spatial technology. The history of place as a philosophical concept has always co-existed with the history of human technological communication networks. It is a key contention of this book that GIS and the GeoWeb share binary/digital geo-referrable points as the basic characteristic of the ontological building block of space, which is today's dominant paradigm in the Western view of space and place. But the spatial humanist must ask what went before this. This chapter traces the historical background to this building block, and seeks to situate it in reference to the means of mapping that went before.

Place in the Ancient World

The first known map of the world is inscribed on a small clay cuneiform tablet, dating approximately to the sixth century BC. It was discovered in 1881 in Babylon, and now on display in the British Museum (Brotton, 2012: 1). This very small, but very important, map consists of two circles, one inside the other, with eight evenly spaced triangles around the outside. This tablet is the first attempt to represent a terrestrial plan of the known earth (the circles), in the cosmographic context of the universe beyond it (the triangles).

Despite the great historical and cartographic importance of this very early visual representation of worldwide scale, interest in the distinction between the cosmographic and the terrestrial stretches back even earlier. Critical understanding of space and place in both abstract and applied terms was of great interest to the philosophers of the ancient world of Greece and Rome. In ancient Greece, the higher branches of the arts, sciences and philosophy were framed in terms of the quest for abstract geometrical perfection; something that Classically inspired European societies up to and beyond the Renaissance period similarly aspired to. When the Parthenon sculptures came to the British Museum in 1801, Lord Elgin's self-professed aim in doing so was to 'improve the arts of England' through the material presence in the country of these paragons of Classical perfection, and the access this would bring to domestic artists and scholars (St Clair, 1998). These artists and scholars were enjoined to seek the attainment of a spatial ideal that was geometrically balanced, mathematically ordered, and bore no linear relationship to the real (and imperfect) world. This was often related directly to notions of Greek cultural exceptionalism, of the supposed superiority of the 'Greek form' over those produced by other cultures; and, by extension, contemporary societies which took inspiration from Classical Greece.

This hyper-nostalgic view of Greek culture was captured by Percy Gardner, when he wrote of Classical Greek sculpture:

> Many barbarous peoples have [produced sculpture], but Greek sculpture became rapidly progressive because the Greek had within him a natural faculty and taste for art. He could see that the figure he had produced was not really like a human being, so he amended it in the direction of naturalism. And he had within him a love of human beauty and strength and symmetry which made him discontented with mere naturalism. (Gardner, 1917: 419–21)

The implication of the figure produced being 'not really like a human' is that to achieve perfection, a sculptor must not seek to execute his creation by copying the real-world subject from observation in his medium, rather he should seek to interpret his subject into a form that reflects abstract perfection. This dichotomy conveys a fundamental distinction between space and place which are relative, embodied, experienced and subjective, versus space and place which are absolute, immutable, unchanging and unrelated to the physical world – which is inevitably imperfect. It also allows us to establish a concrete link between the theoretical view of place as an abstract idea and place as it is lived in the real world, as Yi Fy-Tuan interpreted it.

The Classical Greeks addressed this philosophical distinction directly in approaches to describing place itself, and we can identify parallel streams of thought in the history of cartography. Claudius Ptolemy (c.100–170 AD), in the introduction to his *Geography*, makes a distinction between *geografia*, which he defines as 'the graphic representation of the known world as a whole, including the things that relate to it as a whole', and *chorografia*, which 'takes single regions separately and deals with them individually, embracing every smallest detail, such as creeks, hamlets, villages, the reaches of rivers and such things' (Ptolemy, n.d.: 1,1).

This distinction between 'geography' and 'chorography' has dominated the history of ideas about abstract space ever since. The implication of the distinction is clear: the 'graphic representation' of the earth as a whole is characterized by measurement through macro representation in the former, whereas the 'smallest' detail is characterized by observation by the individual, by perambulating through the landscape. Ptolemy maintains the emphasis of geography as being beyond the scale of human experience, describing it as:

> the subject of the fairest and most sublime contemplation, to reveal to human perception, by mathematics, the nature of the heavens themselves, because they can be observed revolving around us, and of earth by a figure, because the real earth, being very large and not, like the heavens, encompassing us, cannot be traversed either as a whole or as to all its parts by one and the same man. (Ptolemy, n.d.: 1,1).

Geography, therefore, is the scientific measurement of the world at all scales which lie beyond direct human observation. It follows that it cannot be measured by direct human observation either. In Ptolemaic terms therefore, web mapping and GIS echo geography, whereas spatial narrative approaches are more akin to chorography. Like digital place, geography is a structure of logic, which Ptolemy himself sought to make measurable by establishing a catalogue of coordinates referencing specific locations, based on observations of the lengths and angles of shadows at midday, among other things, using a sundial (γνώμων) (Russo, 2013: 76). Ptolemy's dataset, although beset by inconsistencies and historiographical ambiguities (see Isaksen, 2011), represents the first attempt to establish such a logical structure for the world's Geography using technological means, rather than direct human observation. No original Ptolemaic geographic map survives, rather his coordinate tables seem to have been tools for others to generate maps from. Isaksen (2011) observes that they enable the drawing of both geographic and chorographic maps. However, Ptolemy was able to problematize the distinction in the first place by generating logical data (in this case on shadow lengths, from measuring sundials, or gnomons).

Ptolemy's dichotomy between geography and chorography is echoed in later works by thinkers such as René Descartes (1596–1650) and Isaac Newton (1643–1727). Descartes argued that space was an absolute and, along with anything contained within it, could be mapped abstractly on a planar axis, in a manner corresponding with Ptolemy's geography. Space and place, in Descartes' view, was not derived from any human action, perception or intervention, and therefore as well as Ptolemy's geography, it mirrored the kind of absolute detached space to which Greek art, as imagined by Percy Gardener, aspired (see Peuquet, 2002: 21–7). Newtonian physics similarly held that the universe was absolute, non-relative, and governed by fundamental laws that are immutable, and thus observable and predictable. There are therefore broad parallels between the Ptolemaic view of 'geography' and the Cartesian/Newtonian idea of 'mappable' place. Both are equally compatible with the Greek view of abstract perfection, and of a Christian universe created by God from nothing, and of humankind created in God's perfect image. The cultural resonances of the observation and delimitation of space start to become clear.

Case study: chorography and the humanizing of geography

Alternatives to these 'Geographic' views of place emerged in the context of broader intellectual processes from the Renaissance onwards, and one of the most relevant of these for this discussion was a renewed interest in the classical idea of chorography (Olwig, 2011). The post-Tudor period saw a spate of efforts to map, observe and record the landscape and its histories at close quarters, many of which self-described themselves as 'chorographic'. Such efforts were the work of antiquarians, often men of means and education, whose closely observed tracts dealing with topics including the history, topography, natural history, folklore and so on of the nation (or various parts thereof such as counties) were derived from close observation of and perambulation through the landscape. Embodied spatial narratives of the kind described in Chapters 6 and 8 depend heavily on perambulation, and it is no coincidence that this is a style of writing that emerged in this period. Nor was it happenstance that this was also a period of prototypical information overload. Following the invention of printing, and the publication of the Gutenberg Bible some one hundred years previously, movable type and the printing presses were making books more available and accessible than ever. Chorographers and cartographers of the period quickly adopted this new information super-highway for their creative and scholarly purposes; but, as we shall see, it presented them with practical and intellectual problems which shaped the representation of place down to, and into, the digital age.

Antiquarian chorography from the sixteenth to the nineteenth centuries is especially important to the development of landscape phenomenology, the branch of archaeological thought pioneered by Tilley (1994), which holds that a landscape must be experienced directly and personally in order for its meaning to be properly understood. Some have argued that antiquarianism and contemporary phenomenology are linked by 'echoes of choreographic practice' (Gillings, 2011: 60, see also Peterson, 2003), in which structured knowledge about landscape resulted from the combination of then-new methods of formal scientific enquiry, and antiquarian observation derived from perambulation and in-landscape observation. Various examples of the works of well-known antiquarians such as William Stukeley (1687–1765), William Camden (1551–1623) and the High-Church Anglo-Catholic priest Peter Heylyn (1600–1662), exhibit a nascent tension between their concern to record the landscape of England (and, perhaps, that landscape as an emblem of Englishness), and the constraints of the medium of print, which both enabled and restricted the writing of these new narratives through the publishing process.

Partly, this was because of the practical problems of reproducing large-scale maps with the limited capacities of the seventeenth- and eighteenth-century printing press (see below). But also, the wider circulation of these works began to accentuate the separation between author and reader, a *topos* important in later traditions of literary cartography, as will be seen in Chapter 4. The dislocatedness of the relationship between the author and the reader, so obvious in today's world, had a different resonance in the geographies of reading (see his period. While describing individual 'readerly' experiences of the seventeenth and eighteenth centuries is impossible today, it is surely not placing too much burden on the available evidence to suggest that the chorographers popularized the subjective observation of place, at least as far as the literate sections of society were concerned (which is admittedly an important caveat). It was not just the *content* of chorographic works that provide the origins of threads through later spatial humanities practices including (but not limited to – see below) landscape archaeology run (see Shanks and Witmore, 2010), but also that they anticipated ways in which collective spatial intelligence would be shaped and formed by communication media for centuries to come.

This view is supported by examination of the chorographers' methodology. Among the most notable of the early antiquarians was John Leland (c. 1503–1552), whose 1549 *Itinerary*, compiled for Henry VIII in Leland's capacity as 'King's Antiquary', was a painstakingly detailed county-by-county record of the realm's natural, topographical and archaeological features. A firm patriot and nationalist (as any employee of Henry VIII would have had be), Leland was particularly interested in visible archaeological remains, a believer of the literal truth of the Arthurian legends, and an advocate of the view that past and present glories were linked. In his 'New Year's Gift' to the King he begins his address to the King with the words 'I truste right shortely so to describe your moste noble reaulme' (Leland, 1907: xxxviii). In the introduction to this 1907 edition of Leland's *Itinerary*, Lucy Toulmin Smith notes of his method: 'His plan seems to have been to note down his facts on the spot, or from various local enquiries; then later, at leisure, he wrote his narrative direct from them, adding in bits from memory occasionally'(xxxviii). Toulmin Smith talks of Leland making a 'skeleton page' of names of towns, where the detail was sometimes – but not always – filled in later. Such construction of narrative from memory begins to sound very like 'spatial narrative' of the kind described by Bodenhamer et al. (2015), by Lock and Pouncett (2017), and by others who have sought to define 'deep mapping' for the spatial humanities (see Chapter 1).

At the same time, the chorographers used the platform which print gave them to imbue chorography with a fiercely nationalistic character, with Camden asserting in his *Remaines concerning Britaine* that the inhabitants of Britain are 'if answerable worth, if not surpassing, yet equalling the most excellent inhabitnats of the earth, both in the endowments of mind, lineaments of body, and their deportment in both peace and warre' (Camden, 1623: 9).

The socio-religious perspective, enculturing the physical landscape and its antiquarian remains, and broadcasting nationalistic propaganda at the same time, ran deeper, as the authors themselves made clear. They embraced Biblical teaching about the irrelative centrality of humankind in the universe – in the *Remaines*, Camden describes 'Man' as 'being the very image of God, and a petty world within himself' (Camden, 1623: 2); and many were churchmen. However, it is also certainly true that recording natural phenomena, and distributing the outcomes of that process, fostered a spatial identity that was derived from direct observation of the natural environment rather than religious revelation, and that this encouraged relativistic and nationalistic views of place to gain currency through the wider circulation that movable type allowed.

The close, personal observation of particular elements of the landscape, gleaned through first-hand experience and perambulation, and described verbally. Yet the platforms afforded by movable type and the circulation of ideas in print entailed practical difficulties of cost. In framing and publishing their ideas, the antiquarians encountered the novel conundrum of having to factor in the relative costs of visual and textual media. In 1677 Heylyn produced his *Cosmography in four books containing the chorography and history of the whole world and all the principal kingdoms, provinces, seas, and isles thereof* (Figure 2.1), a work of encyclopaedic world geography describing the four quarters of the world. In this, he explicitly rejects visual maps in favour of the textual description – on grounds of cost. This was a problem with which the nature of the subject matter presented the medium, and Heylyn was concerned that adding more maps than necessary to his 'Cosmography' would add to the cost of its publication, and thus reduce his readership:

> I did once think of beautifying the Work with as many maps as the several States and Kingdoms which are here described. But on further consideration, how much it would increase the Book both in bulk and price, and consequently of less publik use than I did intend it; I laid by those thoughts, and rested satisfied with the adding of four Maps for the four parts of the World. (Heylyn, 1677 ["To the Reader"])

Similarly, the abridged 1701 edition of Camden's *Britannia* indicates that the original publication suffered from the prohibitive costs of print. In the preface, the editor states:

'The last edition of our Author, Publish'd by the Ingenious Mr. Gibson, met with that Acceptance in the World as might be expected. But it being a very large Volume, and upon account of its Maps and other Sculptures, unavoidably high in its price, it was thought it might be of Publik Use to Publish and Abridgment of this Author' (Camden, 1701 [unfoliated]); in this case without maps.

The nature and limitations of the print medium thus begin to exercise a shaping influence on both the practice and the reception of chorography itself, and thus on the way observed, recorded and represented place was transmissible to the wider reading public.

The emergence of chorography and socially constructed space up to the nineteenth century illustrates that views of space as a social and anthropogenic construct are closely related to technologies which allow faster and cheaper communication. Chorographic antiquarianism thrived in the age of the printing press, when the minute observations of writers such as Leland, Heylyn and Stukeley could be shared, if not to a mass market, then at least a much wider one. Other antiquarians and large-scale cartographers after the Elizabethan period sought to stress the totality of their abstract cartographic coverage, while not addressing the practical limitations as Camden and Heylyn did. For example, *Ellis's English Atlas, or a Compleat Chorography of England and Wales on Fifty Maps,* published in London in 1773, was derived 'From the latest Surveys of the Several Counties' (Ellis, 1773). It seems therefore that 'chorography' came to be seen as detailed description which, above all, was conveyable and communicable *in printed form*, despite its limitations.

Looking back on this period from the perspective of the contemporary spatial humanities, a key development of this period is the emergence of geographic and chorographic knowledge as a means for understanding the distant past. Stukeley, in his *Account of Richard of Cirencester, monk of Westminster and his Works with his Antient Map of Roman Brittain; and the Itinerary Thereof* describes the work (specifically Chapter VI) as 'an invaluable curiosity to the inquirers into *Roman Brittain*, he gives us an exact and copious chorography of the whole island: its boundaries, rivers, mountains, promontories, roads, nations, cities, and towns, in the time of the *Romans*' (Stukeley, 1757: 14–15 – this follows a detailed account of how Stukeley

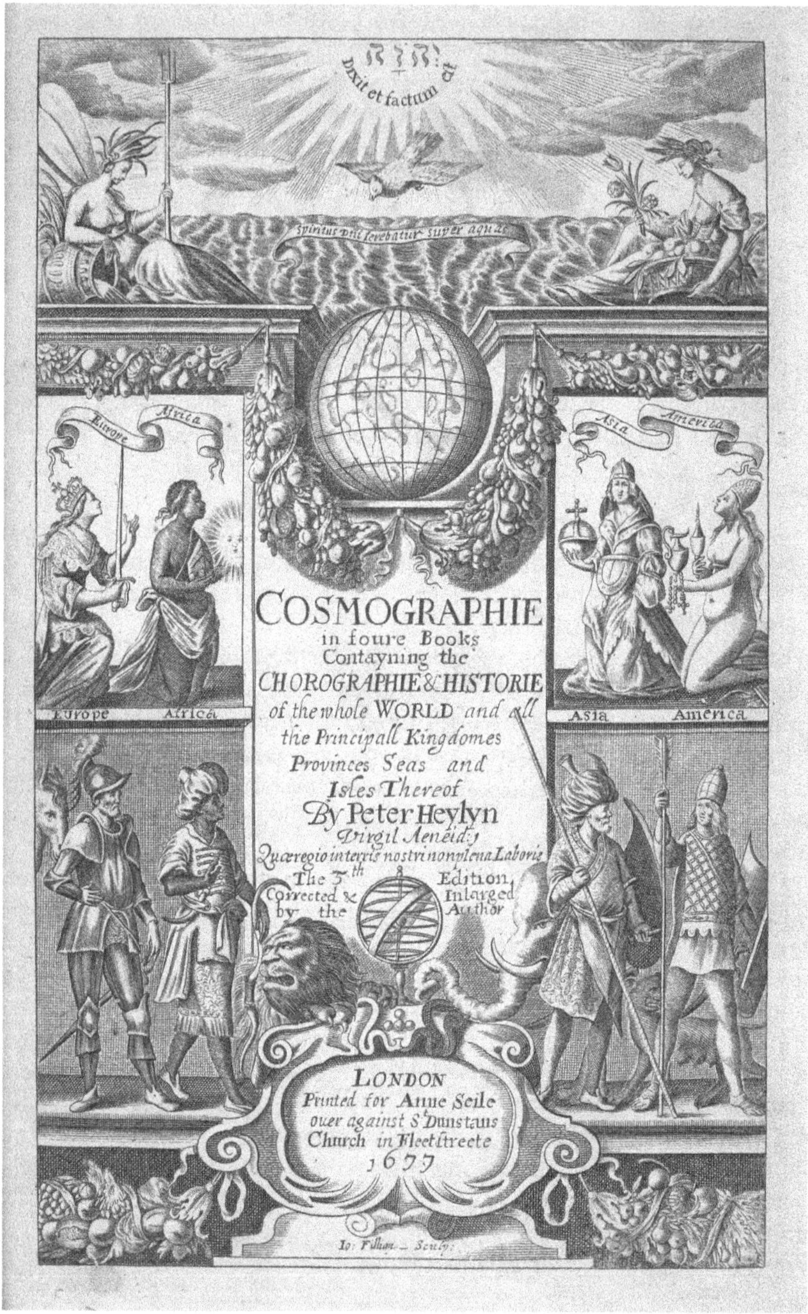

Figure 2.1 Peter Heylyn's Chorography and History of the Whole World: and All the Principal Kingdoms, Provinces, Seas, and the Isles Thereof.

himself came by the manuscript via correspondence with the chief librarian to the Dutch crown).

Abraham Ortelius provides a case in point to the emergence of 'explanatory' spatial narratives which are formed in the present and projected into history. In the introduction to the 1570 edition of the *Theatrum,* which included maps depicting the places and topographies of the Classical and Biblical words, Ortelius refers to the problems of scale and cost alluded to by chorographers such as Heylyn and Camden – and thus explains his notion of selling maps in a book, rather than as individual rolls of vellum – but he goes on to explain the importance of cartography as a narrative vehicle for understanding the past:

> Seeing that as I thinke, there is no man, gentle Reader, but knoweth what, and how great profit the knowledge of Histories doth bring to those which are serious student therein … there is almost no man be it that he have made neuver so little an entrance in to the same … for the understanding of them aright, the knowledge of Geography, which, in that respect is therefore of some – and not without just cause called The eye of History (Ortelius, 1606).

This was a key concept to introduce to a post-Renaissance Europe steeped in the idea of Classical learning as the primary means of understanding and connecting with the ideals of history, as exemplified by the textual testaments of the Greco-Roman world. The idea that a 'knowledge of Geography [is] the eye of history' is a departure from the mindset that in order to understand the Greek world, one must read (for example) Thucydides or Herodotus; or that the key to understanding the Roman world was to read Tacitus or Polybius. Rather, for a true understanding of the narratives these authors describe in their extant texts, one must also understand the geographical setting in which those narratives happened. To this end, Ortelius, through his invention of the atlas, was facilitating the transfer of spatial data and spatial information to provide the necessary framing for a full understanding of the past. While mapping technologies and the visual traditions of cartography have always conditioned our views of the world in the past and the present, Ortelius's 'eye of history' documents a critical appreciation of this fact emerging over four centuries ago. Camden acknowledges this explicitly as a personal debt to Ortelius:

> That excellent reviver of antient geography *Abraham Ortelius*, was extremely urgent with me thirty years ago to illustrate the antient state of my native country of Britain, or, in other words, that I should restore antiquity to Britain and Britain to its antiquity, give antient affairs a new air … and reinstate truth to our histories, from which it had been banished either by the confidence of writers or the credulity of the vulgar' (Camden, 1806).

These writers were articulating a new view of the world, and at the same time understanding its politics and society (and the histories of politics and society) through mapping. As the geographer John Pickles has observed:

> Mapping and cartography – the drawing of lines and the bounding of objects – have been at the heart of … consciousness. Mapping technologies and practices have been crucial to the emergence of modern 'views of the world', Enlightenment sensibilities, and contemporary modernities. The world has literally been made, domesticated and ordered by drawing lines, distinctions, taxonomies and hierarchies: Europe and its others, West and non-West, or people with history and people without. (Pickles, 2004: 4)

This kind of 'view of the world' pertaining to maps, is based upon principles of empirical Cartesian accuracy, and in many cases, as Pickles suggests, these form the basis of explanation through cartography.

Aspiration to such scientific rigour was a significant feature of this period, and it contrasts markedly with the period before. One of the defining features of the so-called 'dark age' in Great Britain, between the end of the Roman occupation and medieval period is the absence of any mapping of the island. The mid-fourteenth-century 'Gough Map' is a testament to the emerging connectivity between centres of population at this time (Lilley, Lloyd, & Campbell, 2009). Later maps, such as Ellis's (see above) focus on the intricate, country-wide network of 'post roads' which had been established between settlements. Such maps convey a narrative of the emerging importance of communication by road, and they are borne out in the medium of the map through the practice of representing visual connections. This contrasts with the explanatory narratives of the medieval cartography which went before, which was not so constrained, and was concerned with the conveyance of global truth, and spatio-temporal narratives about the world, the place of humanity within it, and the spiritual and political truths it contained (Peuquet, 2002: 149). The placing of Jerusalem at the heart of the medieval *Mappa Mundae* for example, spoke not to any geographical or perspectival truth, but to the 'truth' of Jerusalem's centrality to Christian consciousness (Alexander, 1997).

However exploration, and the physical movement of peoples in the process of colonization, meant that spatial information, gathered first hand, could be documented and transmitted to others (in the domestic markets for maps and atlases); and in the process it became increasingly vital for both political and navigational purposes. Europeans of the sixteenth century could trace the outlines of distant, unseen and unexperienced continents in Ortelius's *Theatrum Orbis Terrarum* – the world's first atlas in book form – through the projection system of Gerhardus Mercator, which allowed the curved Earth to be represented on a flat sheet of paper (and this printed, re-printed and sold); and form understandings of these places without ever experiencing them first-hand.

The chorographic antiquarians of the seventeenth and eighteenth centuries drove new perspectives derived from description and documentation of the landscape and its topologies, and these are demonstrable in the individual cases described above. They also drove new perspectives collectively through mass communication, establishing a 'cartographically and chorographically shaped consciousness of national power' (Helgerson, 1986: 52). This process of 'collectivizing place', and the powers behind place, was, at least in part, driven by the affordances of the period's communication technologies, i,e. movable type and engraved maps; and the etymology of the word *chorography* itself hints at the documentary and communicative aspect of the practice; the activity of *writing* or *drawing* about a region (*khōros*). However, ironically enough, this period presaged a later one, driven by the advancing pace of the communicative technologies through which place was recorded and communicated, defined not by the authorial and authoritative voice of individuals with access to the mysteries of the printing press, but rather though the day-to-day interactions of millions of individuals.

Socially constructed place

The period of the 'long eighteenth century' (1685–1815) and its aftermath brought radical changes in society, culture and intellectual outlook. Whereas chorographers had operated firmly within the Christian frameworks of universal abstraction, while imbuing it with a muscular nationalistic flair, post-Darwinian thought challenged this view. The Industrial Revolution and urbanization, and associated upheavals, meant that ideas of space, place and fixity – in both a geographical and a social sense – were challenged. In 1859, Darwin's theory of evolution further subverted the notion that humanity occupied a fixed place in a divinely ordained order of being, suggesting the alternative view that humans defined themselves physically through responses to their needs and environment – the principle of genetic mutation and modification. It is perhaps unsurprising therefore that Newtonian and Cartesian orthodoxies about fixed and non-relative place found themselves similarly usurped by ideas of place which held it to be socially constructed, a product not of the laws of nature and the universe (or God), but of the views and experiences of human beings.

In the nineteenth and twentieth centuries, Marxism became highly influential in the view that space was formed by human perception and interaction rather than perfect abstraction, viewing it as a collectivist construct, constantly made and re-made through conflicting collective social pressures. Henri Lefebvre, one of the most influential drivers of the idea that place is constructed by and through social relations rather than abstract physical laws, expressed this in the context of the geographies of urbanism, writing of 'the existence, from the sixteenth century onwards, of a unitary code or common language of the city' (Lefebvre 1991). In 1957, *The Naked City*, a conceptual map of Paris was published by the *Mouvement Internationale pour un*

Bauhaus Imaginist, affiliates of both the London Psychogeographic Society and the *Internationale situationniste* movement. *The Naked City* deconstructed the municipally produced *Plan de Paris*, a regular geometrically accurate and 'official' cartographic record of the cityscape, and re-imagined it according to how easily the different districts represented could be accessed from each other on the ground. It thus explicitly rejected positivist geography, and stressed an intangible 'unity of atmosphere' as the principal determinant of human movement through the urban cityscape, rather than the cold, quantitative and deterministic layout of planned space (McDonough, 1994).

Collectivizing place at the speed of light

The thread that emerges from this account is that the technologies and philosophies of communication have always been linked to human sense of place. Indeed there is plenty of reason to support the post-Marxist view that faster and cheaper communications were instrumental in enabling the idea of place as a social construct, as articulated by Massey and Lefebvre. In this context, the GeoWeb represents an exponential acceleration of this process, enabling the sharing of individual spaces and places, expressed according to defined standards (such as the WGS84 grid) at the speed of light. As pointed out in Chapter 1, the GeoWeb is a subset of the World Wide Web which acquired a special designation as a result of the special properties of WWW-enabled geographic data, and the particular types of methodology needed to deal with them. However, the WWW on which the GeoWeb itself is based is itself a space-collapsing technology. The GeoWeb infrastructure, and methods for creating and sharing maps is a spatially specific refinement of this.

In the context of the present, spatial humanists are developing increasingly sophisticated ways of tailoring and harnessing these technologies for the creation and expression of spatial narrative, through coding, the building of web platforms, the use of 'reader-generated' maps using mobile technology (Cooper & Priestnall, 2011), the creation of Volunteered Geographic Information (VGI) (Goodchild, 2007) and the development of specialized gazetteers based on places, or events, or time periods, to represent humanities material (Mostern & Johnson, 2008). However, most discussions which address the mediating nature of technology as an enabler of spatial humanities research, even those which foreground the ontological complexities of digital place in the humanities (e.g. Luchetta, 2017), tend to reduce the medium of digital cartographic representation to that of the background map, taking the form of contemporary Cartesian web platforms such as Google Earth and OpenStreetMap. Such platforms give a familiar (today ubiquitously so) 'God's-eye' view of the world which can be panned, zoomed, searched, pinned, interacted with and shared in multiple ways. While the Geoweb of Google Earth, Yahoo maps, Bing, OSM etc, differ from GIS in terms of their functions (the former being representational, the latter analytical), both contribute to our affective sense of place through their use of binary, digital,

Cartesian data, almost always in the Mercator projection system (which itself has considerable interpretational and presentational limits). This is a way of representing place which has evolved to serve the connective infrastructures of the Internet. If the spatial humanities have come to rely on geospatial technologies and the GeoWeb as a medium, then it follows that they have also come to rely on – and be shaped by – the Internet's connective infrastructures.

In an elegantly ironic fork of the historical narrative of the time, these infrastructures emerged, in a comprehensively Lefebvrian manner, from the conflicts of the twentieth century, mainly the events and processes in the decades following World War II. This is a period of particular interest for the technological compression of space, because it saw both an acceleration of social change which strengthened the perceived disconnect between a pre-war society which changed out of all recognition between 1939 and 1945 and its sense of place, and the technologies which disconnected social and physical place. And as discussed further in Chapter 4, it triggered a deluge of academic scientific research output and scholarly communication, leading – indeed forcing – the computer science community at the time to address how volumes of text that were too large for any single human being to digest could be stored, structured and retrieved outside the constraints of physical place (Bush, 1945). Finally, it triggered an arms race which created a need for large-scale computational power for tasks such as anti-aircraft ballistics and the thermodynamic modelling of hydrogen bomb explosions (Dyson, 2012), tasks that were similarly beyond the capacity of any individual human.

The Soviet Union's launch of the Sputnik satellite in October 1957 precipitated a state of national crisis in the United States, leading to a suite of research projects which the US Advanced Research Projects Agency (ARPA) supported, covering a range of computing, graphics – and communication. Most importantly, this included the establishment of high-speed packet switching networks utilizing existing telephonic communication infrastructure (Baran, 1964). These networks facilitated the high-speed transfer of large amounts of data through networks with a high degree of *redundancy* – that is to say, unlike an analog circuit network which is tied to, and connects, two or more specific places, it could function if a certain number of its links (or places) were removed – for example if one node was obliterated in a nuclear strike. These initiatives had the aim of anticipating any further unexpected disadvantages in the scientific and arms races with the USSR (O'Neill, 1995: 76). But it also brought about a change in the philosophy of computing. Whereas previously computers (of the WWII era) had essentially been machines for performing calculations at speeds and scales beyond even the best human 'computers', the vision ARPA had was to facilitate the use of computing to link machines, people and networks across the continental vastness of North America and internationally (Leiner et al., 2009; Abbate, 2000). Through networked connectivity facilitating resource sharing, communication and collaboration using machines, including 'home consoles', computing would develop new ways in which machines and people could think together over the vast distances of the continental United States.

As the realities of the new network infrastructure settled in, so its possibilities of 'space time compression', to quote Massey (see below), began to dawn. It would transcend existing information infrastructures, leading humans to interact with information in new and unforeseen ways. But as the chorographers of three hundred years earlier struggled with the limitations of print publication, so the twentieth century's most influential computing thinkers realized that this new landscape of information overload would present problems of organization, presentation and practice. Most importantly, even in the earliest phases of networked computing, it was recognized that unconstrained capacity for instant communication over large distances would impact on users' ability to make sense of the data being communicated. In a landmark paper, *The Computer as a Communication Device,* J. C. R. Licklider, Director of ARPA's Information Processing Techniques Office from 1962–4, and Robert W. Taylor, noted:

> There comes a time when the complexity of a communications process exceeds the available resources and the capability to cope with it; and at that point one has to simplify and draw conclusions. It is frightening to realize how early and drastically one does simplify, how prematurely one does conclude, even when the stakes are high and when the transmission facilities and information resources are extraordinary. (Licklider & Taylor, 1968: 27).

Licklider, a behavioural scientist, had a vision that – despite the dangers of oversimplification – the barriers that space and time presented to communication and information could be overcome, and that diminishing the costs of time, capacity and energy that these barriers exacted could lead to new forms of thinking outside the physical limitations of space and place (Licklider, 1960). This vision was realized with the formalization of protocols for transmitting information across networks which removed physical location from the concept of 'addressing': communication of information between machines was instead mapped using numerical locations, not physical ones, replacing it with a distributed network of numerical addresses (Internet Protocols, or IPs), across which information could be packeted and conveyed as separate units from any address to any address, irrespective of physical location (Roberts, 1978).

The concept of a network in which information is routed through a series of nodes as packets, rather than via the conventional circuit model overcame two major problems. First, previous communications networks, most obviously the telephone system, depended on single-line circuit connections between two communicating nodes. A call between two nodes would tie up the connection between the nodes for its duration, closing it to all other traffic. The ARPA model of a seamless network of computing resources could not be realized in such conditions. A model in which information is transmitted between fixed nodes along fixed lines necessarily limits the volume of traffic possible at any one time. A distributed packet network, on the other hand, utilizes all possible routes between all possible nodes, removing any limitation that geographical scale imposes on the network. Second, a circuit network which might be traversed by *n* number of routes passing through a limited number of nodes is vulnerable, and the threat of nuclear war in the 1960s meant that network vulnerability could not be countenanced. One of the concerns of ARPA's Information Processing Techniques Office was how the United States' command and control infrastructures would operate in the event of a surprise nuclear attack knocking out key communication nodes which could delay, or even prevent, the issuing and processing of orders for retaliation (O'Neill, 1995). Such concerns had driven the RAND corporation's earlier work on packet switching, in particular the ideas of RAND's Paul C. Baran.

In 1964, Baran prepared a series of memoranda, *On Distributed Communications*, for the US Air Force on the theme of a 'Distributed Adaptive Block Network', which was a digital system for ensuring communications resilience in the face of a catastrophic nuclear exchange. The first of these, 'Introduction to Distributed Communication Networks' compared distributed networks to hierarchical ones, and stressed the importance of redundancy in the former (Baran, 1964). As noted above, redundancy refers to the portions of the network not required for a piece of data to travel along a certain A–B route; and which could therefore be removed without preventing that communication. In Baran's calculation, a relatively low level of redundancy, some three to four times the minimum number of nodes required for any one exchange, would provide a maximum level of resilience in the event of an attack (ibid.). The disconnection of information, and information transfer, from the physicality of distance for a very specific military purpose was, therefore, a key driver of the early Internet.

The purpose of the Internet was to detach resource and information sharing from the constraints of physical and geographical place. This was a purpose ordained collectively by a wide range of agencies, groups, individuals and institutions, although in particular Licklider, in the passages quoted above, articulates it. This was the context into which the World Wide Web was born when Tim Berners-Lee and Robert Cailliau of the CERN facility in Geneva released the HTML protocol as a means of enabling research documents and data to be encoded and shared asynchronously across the same distributed networks (Mowery & Simcoe, 2002: 1378). HTML was derived from hypertext, which enabled users to specify particular sections of text and link them to others. The combination of the 'HyperText Transfer Protocol' (HTTP) and HTML embedded the multimedia characteristics of a physical document in digital form, and allowed it to be transferred from host to host across space, as the Internet had done for programming resources and data packets. This had the effect of subverting the physicality of the document, and rendering its location in physical space irrelevant. Unlike physical documents which had to reside in a physical location such as a library, such documents were only *located* in the sense of being a digital array in a random access memory, which could physically reside, and be accessed from, anywhere. In some ways, this fulfilled the Internet's original vision. As Licklider put it in 1965, '[w]e need to substitute for the book a device that will make it easy to transmit information without transporting material' (1965: 5). In practice, doing so massively disrupted human perceptions of space and place, probably more so than any other in the history of technology.

Place as a social construct

A broad historical spectrum thus links the early descriptions of place in the context of the printing press with the seamless representation of place on the GeoWeb. This is a spectrum in which improvements in communication technologies, including the telephone and mass printing, led gradually over a long period of time, to a massively increased capacity for people (and cultures) to broadcast and replicate their identities. In the age of mass information infrastructures, this profoundly affected political and cultural orders. In the famous example described in her 1991 paper, *A Global Sense of Place* (1991), the feminist geographer Doreen Massey articulates the range of different identities visible in a walk down Kilburn High Street in London, her own local area. Here (she says) one can detect the presence of Irish nationalism, immigration form the Middle East, food from India and so on. Beginning from Marx's aphorism that technology itself was responsible for 'the annihilation of space and time', networks whose edges had previously been defined by the cost/benefit laws of physical geography were now being transcended by instant communication capabilities. Massey called this 'space-time compression', a phenomenon which she defined as 'movement and communication across space, to the geographical stretching-out of social relations, and to our experience of all this' (Massey, 1991: 24) As the spatial footprint of communications available to certain individuals grew, at the same time, the time and effort needed to enact communication shrank.

In a shrinking world, such identities become increasingly detached from the locations in which they originated, and thus they become increasingly interchangeable across different locations. Massey argues that the fluidity of movement brought about by increased global mobility (both informational and physical) reinforced inequalities and imbalances, as space and time was 'compressed' for those with access to resources and capital in way that it was not for those without such access: 'Boeing has brought … people together. But what about those they fly over, on their islands five miles below? How has the mighty 747 brought them greater communion with those whose shores are washed by the same water?' (ibid.: 25).

In the same way, Massey argues, place is represented and replicated across space and time in the information age: 'If it is now recognised that people have multiple identities then the same point can be made in relation to places.

Moreover, such multiple identities can either be a source of richness or a source of conflict, or both' (ibid.: 28). This was a highly prescient statement: present-day social media enables and encourages the creation of the kind of multiple personas (for example social media profiles), which link, publicize, define, reaffirm and challenge the identities of the individuals they represent, while subverting abstract and physical notions of geography and togetherness. This reflects the kind of world which Massey saw emerging in the early 1990s, shortly after the World Wide Web's creation at CERN in 1989, yet was far from reaching any point of maturity at that stage. This allows us to establish a clear spectrum between the observation of place, the documentation of place, the transmission of that documentation, and the effect that that transmission has. This is a spectrum that was responsible for the emergence of the GeoWeb.

Place and communication

The significance for place as a concept of instant communication across large distances attracted the attention of communication theorists at a fairly early stage of the Internet's development, and around the time of the inception of the WWW. In particular, the idea that places have multiple identities expressed through, and contextualized by, individual points of communication, and that these distinctions are blurred by electronic communication, was argued by the communications theorist Joshua Meyrowitz in his 1986 work, *No Sense of Place: The Impact of Electronic Media on Social Behavior*. In this, he stresses the changes in communication which are bought about when the electronic medium physical location from the discourse, and highlights its role in separating physical and social place: 'When we communicate through telephone, radio, television or computer, where we are physically no longer determines where and who we are socially' (Meyrowitz, 1986: 115).

Writing less than 20 years after the invention of the packet-switching Internet, Meyrowitz contends that a narrative description of an event, object, person, experience etc. has certain attributes in the spatial intimacy of the face-to-face conversation. This is a highly specific and limited physical context which implies colocation of the interlocutors; but that the same interaction takes on a different meaning if the context of its reception – its audience – is blended across multiple parts of an electronic communication environment. This may be seen as a consequence of Massey's concept of space-time compression, except whereas Massey is describing the imbalances of capital and its influence on the collective human definition of place, Meyrowitz is concerned with how the interactions themselves change as a result of their increased spatial extent and frequency, and of the reduced time envelope in which they occur.

From WWW to GeoWeb

In tandem with disrupting the dynamics of social communication as described by Meyrowitz, the WWW drove an exponential increase in the uptake of the

Internet. When Mosaic, the first browser which allowed access to the WWW was launched in 1993, it increased Internet traffic by a factor of 3416 (Mowery & Simcoe, 2002: 1378), and this increase continued steadily for the next decade and a half. At the same time, protocols and interfaces were developed which allowed users to contribute data to the WWW, rather than simply consume it as passive readers, so-called Web 2.0 (O'Reilly, 2005).

The combination of vastly increased traffic with the emergence of a hyperlinked web that was fashioned by its users created a new problem: should spatial information, with its particular relationship through geometry to the physical world (and existing frameworks for expressing that geometry systematically, such as the WGS84 grid system – see Chapter 1); and the long human history of representing and describing place in visual and textual form, as the discussion above of chorography suggests, be treated differently from text and image in the HTML and HTTP infrastructure? Should digital maps, and elements of digital data contributed by users which contain spatial and/or georeferenced information, be handled simply as HTML text or embedded imagery? Or was there any way in which the affordances of digital mapping, which had been around since the 1950s (Farman, 2010), could be combined with IP addressing and packet switching technologies in order to manage that data? It soon became clear that the answer was yes, and in 1996, the Multimap and Mapquest corporations launched their online mapping services, providing cartographically visualized databases covering large areas. This represented the moment when *web mapping* first took off (Graham, 2009: 423), and in the process can be said to mark the beginning of the GeoWeb. In 2005, Google launched Google Earth, which further revolutionized collective access to spatial data by linking it to Google search results, and also by visually embedding it with the physical world by overlaying it on actual satellite imagery (Goodchild et al., 2012).

The most significant thing about these innovations is that they introduced the concept and grammars of spatial mark to the seamless and lossless communication of the digital world. In 2004, Google acquired Keyhole, a satellite imagery firm which had produced an XML structure for identifying individual decimal coordinates on its images. 'Keyhole Markup Language' (KML) was implemented as the standard for Google Earth, and was adopted as an open standard by the Open Geospatial Consortium (Fink, 2011: 6). At the core of KML is its geometry attribute, which uses un-projected WGS84 latitude and longitude coordinates to georeference information contained elsewhere within the object's XML tree. This is also used to construct the three main kinds of vector geometry used in GIS: points, lines and polygons (see Lock, 2010). With the traction and impetus of burgeoning WWW and Internet usage in this period, and its branding with the Google imprimatur, the KML XML standard became the lingua franca of place on the WWW. The representation of places in georeferenced databases with common georeferenced frameworks, most notably KML, not only allowed users to have instant access to it, but at the same time it began to alter the behaviour of the users themselves (Zook & Graham, 2007). The visibility of an organization in terms of its web presence, measured in terms of its Google ranking, determines its visibility in KML-based web mapping services, and this drives more users conducting searches to that organization (ibid.). Place becomes collectivized, its significance crowdsourced.

The primacy of online visibility and significance further subverted the notion of physical place, and stimulated new ways of approaching social analysis, which were likewise based on significance and connections rather than physical geography. Network theory, for example, emerged as a means of interrogating and analysing social and cultural phenomena through the interaction of discrete entities, defined as nodes, and factors connecting them, defined as edges. Nodes and edges could be configured to give power to some individuals and sections in the network, and to deny it to others (Castells, 2011). Nodes and edges also became a useful way of analysing the internal dynamics of particular kinds of networks independently of physical distance, such road and rail systems, telephony networks, trading links etc.

The concept of networks also attracted the attention of digital humanists, who used the dynamics of nodes and edges to explore citation patterns, cross-references in historical documents, and the iterations of historical scientific discovery (Graham, Milligan, & Weingart, 2015). The rise of the 'networked' and 'information' societies recalibrated social and historical concepts of power and social change with similar concepts of networks as their focus. Most notably Manuel Castells observed a theory of 'networked power', which explained the twentieth century phenomena of globalization in terms of the way in which 'networked actors' gain power through the way they exploit the membership of networks, and the accumulation of information through those networks: 'communication networks are the fundamental networks of power making in society' (Castells, 2011). This switches the focus of networks from being tools to explore passive data, such as the recurrence of names across historical sources, to a means of exploring dynamism and identifying patterns that are hypothetical and future-facing, and can at best only be partially predicted. As will become clear in future chapters, GeoWeb technologies represent a very particular kind of network, which cross both the exploratory and predictive senses identified here

Conclusions

The spatial turn was a result of developments in spatial technology – GIS, web mapping platforms, spatial data and infrastructure, digital, gazetteers and mobile devices; however thinking discursively about place is a much older field. This chapter has suggested that at certain key stages – the development of the printing press and movable type, the reconfiguration of place as a social rather than an abstract or divine concept, the emergence of the Internet and subsequently the WWW – condition, and are conditioned by, the way place is communicated. Sometimes humans conform to the limitations of the medium, sometimes they react against it. The ancient Greeks articulated a dichotomy between chorographical observation derived from direct experience, and geographical mapping that derives from established quantitative relationships between points on the Earth's surface. John Leland and those who came after him in the tradition of antiquarian chorography developed the former into a fine literary art, but also expanded the epistemological boundaries of chorography to include large-scale mapping of particular regions in fine detail. As well as testing the limits of what Ptolemy would have thought of as chorography, the chorographers also found themselves testing the limits of the print medium for publishing and transmitting their analyses of the English landscape. This is a pattern that persisted, and the present-day spatial humanities is the body of scholarship which finds itself tasked with critical analysis of this process.

The Elizabethan chorographers, and before them the philosophers of antiquity, presaged an age in which the Internet/WWW/Geoweb's frameworks for (a) documenting and (b) mass-communicating place represents only *one* way in which place can be observed and communicated, and that these in fact fit into a long spectrum of social-technological change, which has had a profound effect on human culture. As with Massey's space-time compression, cheaper and more accessible long-distance travel, telephony and, eventually, electronic messaging, improved communications, commerce and globalization had the effect of 'shrinking' the world, and thus redefining people's relationship with space and place. In the context of a field of academic geography that was already questioning the validity of fixed boundaries and quantitative structures to define what place is (see Cresswell, 2015; Ballatore, 2016), this situated the Internet and WWW, and emergent geotechnologies such as GPS, at the middle of debates on how technology was coming to shrink the world, and led to an early demonstration of Castell's theory of networked power, as some nodes/actors – individuals, corporations, organizations and countries – became powerful, and others did not.

It was the introduction of the GeoWeb, and especially of the standardized geometries of spatial annotation and documentation embodied by KML, which bought about a historic impact of communication technologies on the collective, social construction of place. Ptolemy's chorography framed space at the human scale, as distinct from the sublime abstraction of idealized geography. The Elizabethan antiquarians developed and promoted the subjective description of landscape, and their propagation through print media – and experienced its limitations. The social construction of space in the Marxist, post-Darwinian world was a philosophical, and politicized, extension of this choreographic practice. These practices of observing and recording place shared a multivocal subjectivity: by expressing spatial observation through word or through image, they put forward a statement that was susceptible to reading, analysis, interpretation, contestation and refutation: this is the core of Marxist ideology of collectivized space.

The GeoWeb, however, through the application of geographic XML standards privileges instant, frictionless communication, and the absolute and instant replication of platial or spatial documentation anywhere else. With the ubiquity of the WWW, especially the mobile WWW, in developed societies, the GeoWeb has thus – for the first time – collapsed the concepts of chorography and geography together. It enables observation of space and place at human scale and its instant sharing, but its subjectivity is constrained by the requirement to transmogrify it before anything else into the KML communication standard.

References

Abbate, J.(2000). *Inventing the Internet*. MIT Press.

Alexander, P. S. (1997). Jerusalem as the Omphalos of the world : On the history of a geographical concept. *Judaism, Spring*, 147–158.

Ballatore, A. (2016). Prolegomena for an ontology of place. In H. Onsrud & W. Khun (eds.), *Advancing Geographic Information Science: the past and next twenty years* (pp. 91–103). GSDI Association Press.

Baran, P. (1964). On distributed communications: Introduction to distributed communication networks. Available online: https://www.rand.org/content/dam/rand/pubs/research_memoranda/2006/RM3420.pdf (retrieved 25 July 2018).

Bodenhamer, D. J., Corrigan, J., & Harris, T. M. (2010). *The spatial humanities: GIS and the future of humanities scholarship.* Indiana University Press.

Bodenhamer, D. J., Corrigan, J., and Harris, T. (eds.) (2015)*Deep maps and spatial narratives.* Indiana University Press.

Brotton, J. (2012). *A history of the world in twelve maps.* London: Allen Lane.

Bush, V. (1945). As we may think. *Atlantic Monthly*, (July 1945), 112–124.

Camden, W. (1623). *Remaines concerning Brittaine, but especially England, and the inhabitants thereof.* London: Symon Watterson.

Camden, W. (1701). *Camden's Britannia abridg'd; with Improvements and Continuations to this present Time.* London: Joseph Wild.

Camden, W. (1806). *Britannia, or, A chorographical description of the flourishing kingdoms of England, Scotland, and Ireland, and the islands adjacent; from the earliest antiquity / by William Camden ; translated from the edition published by the author in MDCVII, enlarged* . (R. Gough, ed.). Printed for J. Stockdale by J. Nichols and Son.

Castells, M.(2011). A network theory of power. *International Journal of Communication*, 5, 773–787.

Cooper, D. & Priestnall, G. (2011). The processual intertextuality of literary cartographies: critical and digital practices. *The Cartographic Journal*, 48(4), 250–262.

Cresswell, T. (2015). *Place: an introduction.* John Wiley & Sons.

Dyson, G. (2012). *Turing's cathedral: the origins of the digital universe.* Pantheon.

Ellis, J. (1776)*Ellis's English Atlas or a Compleat Chorography of England and Wales in Fifty Maps Containing More Particulars Than Any Other Collection of the Same Kind.* London: Robert Sayer and Carington Bowles.

Farman, J. (2010). Mapping the digital empire: Google Earth and the process of postmodern cartography. *New Media Society*, 12(6), 869–888.

Fink, C. (2011). Mapping together: On collaborative implicit cartographies, their discourses and space construction. *Journal for Theoretical Cartography*, 4, 1–14.

Gardner, P. (1917). Idealism in Greek art. *The Art World*, 1(6), 419.

Gillings, M. (2011). Chorography, phenomenology and the antiquarian tradition. *Cambridge Archaeological Journal*, 21(1), 53–64.
Goodchild, M. F. (2007). Editorial: citizens as voluntary sensors : Spatial data infrastructure in the World of Web 2.0. *International Journal*, 2,24–32.
Goodchild, M. F., Guo, H., Annoni, A., Bian, L., de Bie, K., Campbell, F., … Woodgate, P.(2012). Next-generation digital earth. *Proceedings of the National Academy of Sciences of the United States of America*, 109(28), 11088–11094.
Graham, M. (2009). Neogeography and the palimpsests of place: Web 2.0 and the construction of a virtual Earth. *Tijdschrift Voor Economische En Sociale Geografie*, 101(4), 422–436.
Graham, S., Milligan, I., & Weingart, S. (2015). *Exploring big historical data: The historian's macroscope.* World Scientific Publishing Co Inc.
Helgerson, R. (1986). The land speaks: cartography , chorography , and subversion in Renaissance England. *Representations*, 16(16), 50–85.
Heylyn, P. (1677). *Cosmographie in foure Books Contayning the CHOROGRAPHIE & HISTORIE of the whole WORLD and all the Principall Kingdomes Provinces, Seas and the Isles Thereof.* London: Anne Seile.
Hones, S. (2008). Text as it happens: literary geography. *Geography Compass*, 2(5), 1301–1317.
Isaksen, L. (2011). Lines, damned lines and statistics: unearthing structure in Ptolemy's Geographia. *E-Perimetron*, 6(4), 254–270.
Lefebvre, Henri, tr. by Donald Nicholson-Smith (1991)*The production of space.* Oxford: Blackwell.
Leiner, B. M., Cerf, V. G., Clark, D. D., Kahn, R. E., Kleinrock, L., Lynch, D. C., … Wolff, S. (2009). A brief history of the Internet. *ACM SIGCOMM Computer Communication Review*, 39(5), 22–31.
Leland, J. (1907). *The Itinerary of John Leland in or about the years 1535–1543.* (L. Toulmin Smith, ed.). London: George Bell and Sons.
Licklider, J. C. R. (1960). Man-Computer Symbiosis. *IRE Transactions on Human Factors in Electronics*, 1(3), 1–14.
Licklider, J. C. R. (1965) *Libraries of the Future*, MIT Press, Cambridge, MA: MIT Press.
Licklider, J. C. R., & Taylor, R. W. (1968). The computer as a communication device. *Science and Technology, 76*(2), 1–3.
Lilley, K. D., Lloyd, C. D., & Campbell, B. M. S. (2009). Mapping the realm: a new look at the Gough Map of Britain (c .1360). *Imago Mundi, 61*(1), 1–28.
Lock, G. (2010). Representations of space and place in the humanities. In D. Bodenhamer, J. Corrigan, & T. M. Harris (eds.), *The spatial humanities: GIS and the future of humanities scholarship*, Indiana University Press, 89–109.
Lock, G., & Pouncett, J. (2017). Spatial thinking in archaeology: Is GIS the answer? *Journal of Archaeological Science*, 1–7.
Luchetta, S. (2017). Exploring the literary map: An analytical review of online literary mapping projects. *Geography Compass*, 11(1).
Massey, D. (1991). A global sense of place. *Marxism Today, 35*, 315–323.
McDonough, T. F. (1994)Situationist space. *October*67, 5–77.
Meyrowitz, J. (1986). *No sense of place: The impact of electronic media on social behavior.* Oxford University Press.
Mostern, R. & Johnson, I. (2008). From named place to naming event: creating gazetteers for history. *International Journal of Geographical Information Science*, 22(10), 1091–1108.

Mowery, D. C. & Simcoe, T. (2002). Is the Internet a US invention?—an economic and technological history of computer networking. *Research Policy*, 31(8–9), 1369–1387.

O'Neill, J. E. (1995). The role of ARPA in the development of the ARPANET, 1961–1972. *IEEE Annals of the History of Computing*, 17(4), 76–81.

O'Reilly, T. (2005). What Is Web 2.0. Available online: https://www.oreilly.com/pub/a/web2/archive/what-is-web-20.html (retrieved 22 July 2018)

Olwig, K. R. (2011). Choros, chora and the question of landscape. In S. Daniels, D. DeLyser, N. J. Entrikin, & D. Richardson (eds.), *Envisoning landscapes, making worlds: geography and the humanities* (pp. 44–54). Abingdon, UK: Routledge.

Ortelius, A. (1606). *Theatrvm Orbis Terrarvm : The Theatre of the Whole World : set forth by that Excellent Geographer Abraham Ortelius.* London: John Norton.

Peterson, R. (2003). William Stukeley: an eighteenth-century phenomenologist? *Antiquity*, 77, 394–400.

Peuquet, D. J. (2002). *Representations of space and time.* Guilford Press.

Pickles, J. (2004). *A history of spaces: cartographic reason, mapping, and the geo-coded world.* Psychology Press.

Ptolemy. (n.d.). *Geography.*

Roberts, L. G. (1978). The evolution of packet switching., *Proceedings of the IEEE*, 66(11), 1307–1313.

Russo, L. (2013). Ptolemy's longitudes and Eratosthenes' measurement of the earth's circumference. *Mathematics and Mechanics of Complex Systems*, 1(1), 67–79.

Shanks, M., & Whitmore, C. (2010). Echoes across the past : Chorography and topography in antiquarian engagements with place Echoes across the Past Chorography and topography in antiquarian engagements with place, *Performance Research*, 15(4), 97–106.

St Clair, W. (1998). *Lord Elgin and the marbles.* Oxford University Press.

Stukeley, W. (1757). *An Account of Richard of Cirencester, Monk of Westminster, and of his Works: with his Antient Map of Roman Britain; and the Itinerary thereof.* London: Richard Hett.

Tilley, C. (1994). *A phenomenology of landscape: places, paths, and monuments.* Berg.

Zook, M. A. & Graham, M. (2007). Mapping DigiPlace: geocoded internet data and the representation of place. *Environment and Planning B: Planning and Design*, 34(3), 466–482.

3 The *longue durée* of the spatial humanities

Part II: The case of archaeology

Introduction

The previous chapter offered a brief overview of the long history of the communication of space and place through human discourse, and how they are shaped in the process by theoretical and philosophical constructs. It was suggested, through the windows into the past afforded by the growth of chorographic antiquarianism in the early modern period, and in the twentieth century by the growth of the Internet and WWW, that this is always a process which is inextricably linked to human capacities to communicate, and the media through which they do so. This chapter takes a narrower disciplinary focus by examining spatial technology, spatial thinking and the specific case of archaeology. This is justified for several reasons. Archaeology is the area of the humanities with the longest history of engagement with GIS, a history due largely to its emphasis on quantitative methods, and its data-driven approach (e.g. Lock, 2010). Therefore no study of the spatial humanities can omit a specific consideration of archaeology. In more recent periods, archaeologists have made extensive use of the platforms of the GeoWeb, in both their research and communication (Harris, 2012) The history of quantification and database construction in archaeology, and its history of digital communication, therefore gives it a perspective on the spatial humanities which is driven by both content and by communication.

This is due to the subject matter and the key questions about the past which archaeologists ask, as well as the methodologies they use. Our view of the past, both distant and recent, is fundamentally connected with our spatial identity and spatial narrative (an idea explored in more detail in Chapter 6). Communal memory of living in a particular location can foster a sense of communal ownership, sometimes to the extent of excluding others from that environment. The seeds of the kind of nationalism are represented in the spatially framed and informed writings of Camden and Leland, as highlighted in the previous chapter. But archaeology, and cognate disciplines of historical anthropology, also frame human social and cultural understanding of the human place within the universe – and critical spatial distinctions, such as that between *geografia* and *chorografia* underpin this. A view of space and place dominated by Newtonian physics, which assumed that all aspects of the universe (and thus space within it) were governed by absolute, quantifiable and observable laws is logically measurable; an abstract entity, static and extrinsic to human action, both containing and constraining human existence (Peuquet, 2002: 21). This form of spatial thinking was eminently compatible with the Renaissance-era intellectual climate, which focused on Platonic and Aristotelian concepts of space such as *chora* and *choragrafia* (the study of peoples and cities), *topos* (topography) and *geografia* (the physical Earth (Cresswell, 2015: 25–30)). By extension, this way of looking at place in the present as abstract and immutable translated logically to a similar view of human progression in the past.

This manifested itself in an historicist view that the human past operated according to immutable and ineffable processes, which are by definition beyond any human understanding or control. Broadly speaking, this view of space, and of the spatial past, dominated most branches of European thought, including archaeology, until the nineteenth and twentieth centuries. However, the processes of urbanization and industrialization of the eighteenth and nineteenth centuries created a climate of intellectual and philosophical, as well as social upheaval. Against this background, a major point of disruption came in 1859, with the publication of *Origin of Species.* The theory of evolution separated the fate of humankind from that of a universe ordered by divine fiat, where 'species were immutable productions, and had been separately created' (Darwin, 1859: 53) placing those species' fate instead in the hands of an unending chain of randomized mutations, conditioned and enabled by survival of the fittest.

Disruptive intellectual events such as Darwin's theory formed the backdrop for later philosophers such as Michel Foucault and Gaston de Bachelard, who laid the groundwork for a conception of place in both the past and the present which was not defined by divine fiat nor by universal mathematical ideals, but constructed by and through human experience of, and action within, the world. This evolution in thinking can be linked to broader themes in human history in the nineteenth century, such as urbanization and the rise of the nation state (helping us in our task of situating the 'spatial' at the heart of the study of human culture and history). Accompanying this was the emergence of an ontological view of place as an entity defined for individual humans by the scope and scale of individual human perception, meaning that places *are* places because they define scopes of human awareness, not because the form part of an ineffable and divinely ordained universe.

GIS and archaeology

GIS entered the arena of archaeological scholarship towards the end of the twentieth century (Lock, 2001), and was swiftly positioned as a methodological nexus, which sat – sometimes uncomfortably – between the practice of archaeology, including the retrieval and analysis of data, and theories of the past, as framed by the epistemological currents described above. Many recognized GIS as a powerful and flexible means to query, analyse, link, interpret and visualize archaeological data, but others regarded it as overly positivist and reductionist, encoding significant assumptions about the past that could not be justified through the discursive interpretation of material evidence on which archaeology had always relied (Wheatley, 2004). As elsewhere in the spatial humanities, GIS acted simultaneously as an enabler of humanistic investigation, and a suppressor of the uncertainty which drives it.

As with other aspects of the spatial turn to which GIS and cognate technologies acted as an accelerant (see Chapter 1), despite criticisms of reductionism, it did not by any means introduce empiricism or positivism to archaeology. Once GIS became available, its adoption, and its peculiar role in the history of archaeological thought, was an inevitable consequence of the fact that understanding the spatial contexts of excavation and finds data is at the core of all archaeological interpretation, both in the field and post-excavation (Lock, 2003: 14–16). More importantly than this, the history of spatial technology and the history of theory in archaeology have a symbiotic relationship, which reflects the distinction between the Newtonian and Darwinian views outlined above.

In Europe and North America, the 1970s saw the emergence of the 'New Archaeology', which held that the past could be studied and reconstructed as a series of processes conforming to universal principles and precepts, rather than the vagaries of individual human agency. The observational, or participatory, context of individual human actors was deemed subservient to 'universal' positivist structures, through which the human past, including distant periods of prehistory, could be understood. The obvious implication of this is that the past can be studied objectively rather than interpretively; and one logical consequence of this is that space and place in the past should be predictable and objectively re-constructible through statistical methods and Euclidean geometry. Central Place Theory, for example, held that the distribution and size of settlements within a given region could be modelled, and therefore predicted, by the systemic characteristics of that region, such as water supply, overland communications, terrain, and so on; an approach which formed the theoretical basis of David Clarke's landmark *Spatial Archaeology*, which appeared in 1977. Tellingly, applications of statistical theory in archaeology of the time, most notably Hodder and Orton's *Spatial Analysis in Archaeology* (1976) saw lack of data, and the possibility of different (human) processes leading to the same statistical results, as a major threat to the future of archaeology, rather than any epistemological problem with the approach itself (Hodder & Orton, 1976: 239).

The early 1980s saw attempts to bring together the traditions of archaeology and geography, where the 'New Geography' was making similar positivist/empiricist claims for the study of spatial structures. The archaeologist Colin Renfew proposed the umbrella term 'Geoarchaeology' for the set of practices which sought to contextualize the physical remains of past societies as elements of spaces with histories of human occupation. Geoarchaeology stressed the relatedness between archaeology, as a discipline whose concern historically was one of time, and geography, whose concern was historically once of space, now that it was accepted that both space *and* time were of importance to both human history *and* the natural environment. In 1983, Renfrew wrote:

> The study of formation processes in relation to landscape features is, of course, a basic concern of the geomorphologist, but only rarely in the past has he or she turned in a concerted way to archaeological sites, and considered them from the same perspective. The archaeologist has recently come to study more closely the way in which human cultural materials are deposited, but rarely integrates these products of human behaviour in any effective way with the relevant geomorphological processes. (Renfrew, 1983: 317).

This focus on methodological crossover between archaeology and geography assumed that both disciplines shared a 'bridging' role between the environment of the human and physical elements (Renfrew, 1983). However, this still carries a basic assumption that space itself was a static backdrop to historical and archaeological processes, albeit one where the distinction between human processes (such as the building of a house) and nonhuman processes (such as the destruction of that house in a hurricane) became theoretically blurred, and defined in terms of common spatial patterns.

The 1980s and 1990s however saw a questioning of this 'processual' approach to archaeology. 'Post-processualism' rejected the notion that the interpretation of the past was extrinsic to human experience, and instead stressed the 'situatedness' of individuals, and the importance of individual experience. In this 'post-processualist' view, the application of GIS which, as noted above, implies an unproblematic association between the place in human record of the real world and place in the vectorised world of the geodatabase, led to a 'tension between the availability of techniques ideally suited to the agenda of the 1970s spatial archaeology and the fact that much of this agenda has been discarded by post-processualists more concerned with a situated contextualized meaning that the identification of general processes' (Conolly & Lake, 2006: 3). This tension remains to this day a critical focus of the adoption of GIS not only in archaeological theory, but in the spatial humanities as well.

Just as thinkers from Foucault to Massey stressed the importance of the social aspect of space rather than its purely physical or binary aspects, so a view emerged in archaeology of space as being produced by social relations and social agency, rather than positivist structures that can be statistically reconstructed (Ashmore, 2002),. This line of reasoning found its apotheosis in archaeological approaches drawing on phenomenology, a branch of philosophical theory rooted in the works of Hurssell and Heidegger (Brück, 2005).

One of the key texts of phenomenology in archaeology, Christopher Tilley's *Paths, places and monuments: A phenomenology of landscape* (1994) states that

> [s]pace can only exist as a set of relations between things or places. In this sense there is no space that is not relational. Space is created by social relations, natural and cultural objects. It is a production, an achievement, rather than an autonomous reality in which things or people are located or 'found'. (Tilley, 1994: 17)

Tilley's ideas as a basis for the formation of embodied spatial narratives are explored further in Chapter 6.

Many have questioned the methodological and theoretical basis of phenomenology (see Barrett & Ko, 2009 and Brück, 2005 for summaries), focusing particularly on its implied claim to be able to reconstruct past human cognitive perceptions of space and place; a claim which is without doubt extremely problematic. However, this emphasis on the role of *social* processes (and phenomena) in the creation (and interpretation) of the material record resonates with many more mainstream branches of contemporary post-processual archaeological theory. Post-processual archaeologists might maintain, for example, that an ethnic Fijian necklace is the result of an entirely unique set of linked social and physical processes which led to its creation, use, modification, and eventual degradation (Gosden & Marshall, 1999: 170). Like all artefacts, the necklace is a 'social object'.

Texts, from the cuneiform writings of ancient Babylonia to modern literary fiction, are also 'social objects', written in a certain social context, and drawing on particular authorial viewpoints, interpretations and knowledge. And digitization, as a link between text and the social affordances of the digital world, can be an immensely powerful tool. As McGann has put it: '[o]ne advantage digitization has over paper-based instruments comes not from the computer's modeling powers, but from its greater capacity for simulating phenomena, in this case bibliographical and socio-textual phenomena' (McGann, 2006). Text thus presents a very different set of requirements, and critical questions, for GIS, and even more to the spatial humanities. This is picked up in greater detail in the next chapter.

Archaeology and neogeography

The phenomenon of neogeography is explored in more detail in Chapter 5 (and indeed an archaeological case study is presented there), but given the discussion above, it is worth reflecting here in more detail on the specific relevance of User Generated Content to archaeology. There can be little doubt that Google Earth, launched in 2005 (having been acquired by Google from Keyhole Inc., where it was previously known as Earth Viewer), was a significant moment for archaeological research and, as explained in Chapter 5, for the relationship between archaeological researchers and the general public. Instant, free access to high quality satellite photography coverage of the Earth's surface revolutionized the public's perceptions of place, 'not as disembodied voyeurs, but as participants in global dialog, represented spatially on the digital map' (Farman, 2010). It ushered in a new age of location-based applications and forced a rethink of how the Internet conveys geospatial data about the human past.

Neogeography is a term which was coined in the 2000s to describe the proliferation of spatial content on the World Wide Web, driven by the launch of corporate products such as Google Earth in that period (Turner, 2006). The essential feature of neogeography is that users can, with mobile devices fitted with GPS sensors, create (and curate) a theoretically unlimited amount of georeferenced data on the fly as they move through the world (see Chapter 5). It therefore represents a technological shift, placing computerized mapping in the hands of the amateur many rather than the expert few. In archaeology, this enables researchers to leverage public participation in exploring large-scale datasets such as satellite imagery. For example it helps archaeologists identify possible previously unknown features in remote regions by reviewing vast quantities of satellites images and reporting any possible sightings (Harris, 2012). As well as opening up some of the geospatial visualization methods familiar to expert GIS practitioners, neogeography led to an exponential proliferation in the sheer quantity of geolocated archaeological data available to the Internet user community (Kansa, Kansa, & Watrall, 2011), and can be seen as one of the early drivers of the Big Data agenda.

As with all other aspects of the spatial humanities, the emergence of neogeography as an important component of archaeological research must be seen in the context of technological development and the way in which communications media enable the communication of place. In this, we can see again the ways in which technological innovation frames intellectual debates, with neogeography impacting intellectual processes in archaeology, much as GIS did 30 years ago, and shaping archaeological thought in subtle and unexpected ways. As with broader applications of neogeography, this context is the development of the interactive WWW so-called Web 2.0. A key volume in this regard is Kansa, Kansa and Watrall's *Archaeology 2.0: New approaches to communication and collaboration*, which sought to frame Web 2.0 as a set of tools for remixing, classifying and enhancing data, as well as generating it through direct observation of the archaeological landscape (Kansa, Kansa, & Watrall, 2011). This volume stresses that Web 2.0 is a relatively easy set of tools for users to employ, requiring little in the way of expertise, but potentially leading to significant ways of public participation in archaeological research (ibid. 3–4).

This analysis sets archaeological use of neogeography firmly on the desktop, in a context of linking documents and information that have been produced elsewhere, rather than in the field. This raises the question as to the level and value of that participation as, in Ian Hodder's famous dictum, 'interpretation occurs at the trowel's edge' (Hodder, 1999). But even at that time, researchers were beginning to assess the potential for mobile computing and real-time participation in archaeology, and platforms which made use of it, such as social media. A contribution to another significant volume of this period, *Archaeology and Digital Communication*, edited by Chiara Bonacchi, for example stresses the importance of emerging networks of Twitter users using hashtags such as #dayofarch plays 'an important role in the tool-kit of the digitally minded archaeologist' (Richardson, 2012). In this regard, technological disruption subverts and blurs institutional and professional boundaries and where, in relation to those boundaries, the research process occurs.

In this vein, engagement with landscape via the capture of data using personal electronic devices provides a significant crossover between archaeological networks on mobile platforms and the work of literary geographers, described in the next chapter. Such scholars and practitioners likewise use mobile infrastructure to engage people with texts written about places in real time and real place. Applications such as that developed by Cooper and Priestnall, for example, explored in more detail in Chapter 6, draw on the technologies and methods of neogeography, underpinned by GPS (including proprietary hardware and software, as well as open source code and standards), but of course are driven by a different set of epistemological drivers than archaeologists' recording sites. Physical interaction with landscape – the fact that you have to be in a particular place to be able to record your location there, or any attribute of that location, such as a photograph – is significant to both archaeology and (embodied) literary cartography. This highlights the need for careful description and contextualization of spatial methodology as it crosses application areas.

Archaeological neogeography, therefore, is a mode of interaction and, possibly, a means of intellectual value creation; but it is not yet a mature set of methods, techniques, software and practices in the same way that GIS is. In fact, it is the *lack* of framework equivalent to 'Critical GIS' for neogeography that has been identified as the reason why its promise for 'democratization' has not been realized (Leszczynski, 2014; Haklay, 2013, see Chapter 5 for full discussion). Neogeography's impact on the spatial humanities over the next few years is likely to be in the form of specific, closely curated, critically grounded and co-produced activities, such as that described by Cooper and Priestnall, and not those which harness 'the power of the crowd', as was the claim of some of its early pioneers, and of current platforms such as OpenStreetMap.

Another significant development in Web geography, which has emerged from a combination of Radio Frequency Identification (RFID) technology and Location Based Services is the so-called Internet of Things (IoT) (Feng Xia et al., 2012). The IoT has seen much commercial uptake in certain sectors where information and objects need to be managed over large geographical spaces, for example in supermarket retail and in large-scale agriculture. The IoT will undoubtedly add to the body of world geo-sensors over the next ten years, and form a rich body of research questions and data for the digital humanities. Already the nature of objects in the IoT is being probelmatized. Alexis Lloyd for example defines IoT 'Things' in terms of *portals, subjects* and *oracles* (see Lloyd, 2013), reflecting their potential for respectively accessing information, being subjects of study, and as sources of information. While the place of the IoT in the spatial humanities is not yet clear, it is certain that it will bring the kinds of transformative challenges in the 2010s that GIS brought in the 1990s and 2000s, requiring responses equivalent to the scope and depth of Critical GIS in that period.

Conclusion

The study of digital mapping and archaeology has many threads. It is a story of the emergence of a lived-in conception of space from positivist, post-Newtonian philosophy; the emergence (and abandonment) of the idea that space is a positive absolute which can be used as a statistical tool to reconstruct past societies; and of the complex nuances of the relationship between space and place and text, in all of the latter's various iterations and taxonomies. Like every other area of the Digital Humanities however, it is mainly a story of the adoption, explication, and critical grounding of technology to understand the human record. Throughout this chapter, I have felt the need to insert the word 'Digital' into the term 'Spatial (Digital) Humanities'. Space (and place) have been objects of study in their own right for centuries, but technologies of representation and analysis have catalysed that in ways that could not have been foreseeable 20 years ago.

References

Ashmore, W. (2002) 'Decisions and dispositions': Socializing spatial archaeology. *American Antiquity,* 104(4), 1172–1183. Available online: https://doi.org/10.1525/aa.2002.104.4.1172

Barrett, J. C. & Ko, I. (2009). A phenomenology of landscape: A crisis in British landscape archaeology? *Journal of Social Archaeology*, 9(3), 275–294.

Bonacchi, C. (2012)*Archaeology and digital communication. Towards strategies of public engagement.* Archetype Publications.

Brück, J. (2005). Experiencing the past? The development of a phenomenological archaeology in British prehistory. *Archaeological Dialogues*, 12(1), 45–72.

Clarke, D. L. (ed.) (1977)*Spatial archaeology.* Academic Press.

Conolly, J. & Lake, M. (2006)*Geographical information systems in archaeology.* Cambridge: Cambridge University Press.

Cresswell, T. (2015)*Place: an introduction.* Chichester: Wiley Blackwell.

Darwin, C. (2009 [1859]). *The Origin of Species by means of natural selection, or the preservation of favoured races in the struggle for life.* Penguin Classics.

Farman, J. (2010). Mapping the digital empire: Google Earth and the process of postmodern cartography. *New Media Society*, 12(6), 869–888. Available online: https://doi.org/10.1177/1461444809350900

Gosden, C. & Marshall, Y. (1999). The cultural biography of objects. *World Archaeology,* 31(2), 169–178. Available online: https://doi.org/10.1080/00438243.1999.9980439

Haklay, M. (2013). Neogeography and the delusion of democratisation. *Environment and Planning A*, 45(1), 55–69. Available online: https://doi.org/10.1068/a45184

Harris, T. M. (2012). Interfacing archaeology and the world of citizen sensors: exploring the impact of neogeography and volunteered geographic information on an authenticated archaeology. *World Archaeology*, 44(4), 580–591.

Hodder, I. (1999). *The archaeological process. An introduction.* Oxford: Blackwell Publishers.

Hodder, I. & Orton, C. (1976)*Spatial analysis in archaeology.* Cambridge: Cambridge University Press.

Kansa, E. C., Kansa, S. W., & Watrall, E. (eds.). (2011). *Archaeology 2.0: New approaches to communication and collaboration.* Los Angeles: Cotsen Institute of Archaeology Press.

Leszczynski, A. (2014). On the neo in neogeography. *Annals of the Association of American Geographers*, 104(1), 60–79. Available online: https://doi.org/10.1080/00045608.2013.846159

Lloyd, A. (2013)If this toaster could talk. Narratives in the age of the smart object. *The Atlantic,* September. Available online: http://www.theatlantic.com/technology/archive/2013/09/if-this-toaster-could-talk/279276/,

Lock, G. (2001). Theorising the practice or practicing the theory: archaeology and GIS. *Archaeologia Polona*, 39,153–164.

Lock, G. (2003)*Using computers in archaeology: towards virtual pasts.* London: Routledge,

Lock, G. (2010). Representations of space and place in the humanities. In D. Bodenhamer, J. Corrigan, & T. M. Harris (eds.), *The spatial humanities: GIS and the future of humanities scholarship* (pp. 89–109). Indiana University Press.

McGann, J. (2006). From text to work: Digital tools and the emergence of the social text. *Romanticism on the Net*, 16(41–42),49–62.

Peuquet, D. J. (2002)*Representations of space and time.* New York: Guilford Press.

Renfrew, C. (1983). Geography, archaeolology and environment. *The Geographical Journal*, 149(3), 316–333.

Richardson, L. (2012). Twitter and archaeology: An archaeological network in 140 characters or less. In C. Bonacchi (ed.), *Archaeology and Digital Communication*, 15–24. London: Archetype Publications.

Tilley, C. (1994). *A phenomenology of landscape: places, paths, and monuments.* Berg.

Turner, A. (2006)*Introduction to neogeography.* O'Reilly Media, Inc.

Wheatley, D. (2004). Making space for an archaeology of place. *Internet Archaeology, 15.*

Xia, F., Yang, L. T., Wang, L., & Vinel, A. (2012). Internet of things . *International Journal of Communication Systems,* 25,1101–1102.

4 Text and place

Introduction: The textuality of space, and the spatiality of text

As discussed in the previous chapter, archaeology was the arena in which the humanities first engaged in a serious way with GIS, accentuating debates about the past that had gone before, and distinctions between spatial theory and spatial practice. This fits in to a wider context of human place-making through communication. The history of the Digital Humanities is heavily dominated by text, and by digital approaches to textual scholarship. The aim of this chapter is to explore how spatial technologies have impacted on that history and ask how their application might lead to common approaches in the areas of spatial narrative and spatial data infrastructures. It is argued that spatial methodologies impact on the reception of text in three broad ways: as means of *communicating*, of *organizing*, and of *reading*.

Text of any kind of genre, form, language or purpose can be used to describe and represent place in myriad ways. However the tools and affordances of representational cartography are ill-equipped to deal with text's social and discursive complexity (Hones, 2008; Rossetto, 2016). As described in Chapter 1, Andrews defines maps as 'a partly schematized graphic representation whose signs are interrelated in ways that resemble, or could reasonably be thought to resemble, the horizontal relations connecting the objects represented' (Andrews, 2009: 1). This representation-focused definition is antithetical to the discursive and descriptive nature of the textual formation of place, be it in fact or fiction. This is one of the key challenges that the emerging field of literary cartography, (formerly literary GIS), has had to contend with. Franco Moretti for example sees the map as an analytical tool through which reflexive meaning can be explored from an empirical starting point: 'Of maps, I mean, not as metaphors, and even as ornaments of discourse, but as analytical tools that direct the text in an unusual way, bringing to light relations that would otherwise remain hidden' (Moretti, 1998: 3). Emphasizing the parallels with archaeology, this contrasts with the position of Hodder and Orton, who declared that maps in archaeological texts could, through formal statistical modelling, become sources of new analysis themselves, rather than being simple illustrations conveying texts' meaning (Hodder & Orton, 1976: 2–3).

As Chapter 3 highlighted, much of the dominant thinking around humanities GIS has occurred within archaeology and landscape studies, and text does not have this same level of historical engagement with GIS. However spatial approaches still matter profoundly to text. This is evident from the discussion in Chapter 2 dealing with the way in which the Elizabethan chorographers, utilizing the then-relatively new media of print and movable type, shaped a new approach to the textual recording and the transmission of place (Olwig, 2011). This in turn tapped into much older historical traditions, dating back at least to Ptolemy, of subjective and observable place (*chorografia*) versus abstractly mappable place (*geografia*); a distinction to which we will return later in this chapter. This hints at some of the complexities which arise from the nexus of GIS and the humanities outside archaeology. Others have problematized this by arguing that archaeology and the broader spatial/digital humanities can learn from one another, emphasizing that both traditions place emphasis on narrative, storytelling, and the affectations of change over time (Earley-Spadoni, 2017).

The key premise of this book is that the spatial turn, and the spatial humanities, are products of ever evolving and ever changing approaches to the communication of place from human to human through ever more sophisticated technological means. It is therefore impossible to detach the spatiality of text and the textuality of space from the WWW and digitalization. Both have comprehensively transformed the way text, and thus textually described place, is read, received, interpreted, used, stored, retrieved and communicated. The purpose of the two innovations behind the WWW, Hypertext, and the Hypertext Transfer Protocol, were to structure and organize text more efficiently, to make it more easily navigable, and thus to increase its value to the reader as a repository of the sorts of knowledge the reader is interested in (Bolter, 1991; Ferdinand, 2000). On the other hand text – in the Western literary tradition – is inherently linear, static and authorial. This linearity, and the pre-existing structure of stylistics and grammar, make it susceptible to processing and markup with schemata such as the Text Encoding Initiative (TEI), which 'collectively develops and maintains a standard for the representation of texts in digital form' ('TEI: Text Encoding Initiative,' n.d.). As far as the transmission of place through text is concerned, the tension lies with the fact that *processing* is not the same as *reading* – yet both structure place and space into meaningful entities. So how does a utilitarian rationale for the computerization of text further the geographic and chorographic concerns of the spatial humanist?

This basic question goes a long way explaining the emergence in the last ten years or so of the field of literary cartography, an area of study which has only quite recently begun to address the implications of digital mapping for the field. Most significantly for the present discussion, the question is provoked at least in part by new forms of reading driven by the hypertextual WWW, and the differences this promulgates between 'writerly' and 'readerly' mapping (Gregory & Cooper, 2010), and the idea of text as a dynamic and processual entity, rather than a static product (McGann, 2006; Saunders, 2016).

Many of the traditions of 'deep mapping' focus on the exploratory and descriptive writings of particular authors, such as William Least Heat-Moon (Gregory-Guider, 2004), who adopt avowedly 'spatial' approaches in their writing, and provoke spatial-literary responses in their readerships. However, a barrier to the methodological exploration any epistemology shared with other areas, such as archaeology, is that textual sources are less aligned with the world of literal, physical or human geography than are physical sites and the landscapes. The lack of physicality and 'mappability' of text is at least partially responsible for the fact that textual mapping studies (with exceptions, such as those referred to above) have – until the last five years or so – engaged comparatively little beyond representational mapping approaches, or straightforwardly illustrative modes of map production and spatial expression (Rossetto, 2013). Such representational linearity – the equation, for example, of representing a list of place names referred to in a text on a map with the construction of a spatial narrative – leads to a black-boxing of the relationship between text and maps, with little attention – as compared to the vast literatures on archaeological GIS – given to theorization beyond linear spatial representation.

Collective text, collective place

As noted in Chapter 2, a key distinction in the history of thought on humanistic place emphasizes the difference between place as a social construct, not merely an abstract one. This reflects the Marxist perspective of key thinkers such as Lefebvre and Massey, and the emergence of so-called hybrid geographies, which reject rigid categorizations of place and space, including qualitative vs. quantitative data and human versus physical geography (Sui & DeLyser, 2012). The rejection of such categorizations necessarily requires some form of qualitative, communicative framework in which place and space are viewed.

Place as a human(ist) object, described, understood, referred to, shared, read and consumed is a collective construct, formed by countless iterations at points of human interaction and communication, and 'remembered' both collectively and individually, employing a broad corpus of what Wertsch describes as the 'cultural tools' of memory (Wertsch, 2008: 121). The most culturally embedded of these tools is text, with its ambiguous relationship between the hyper-empirical practice of information processing, and the discursive practice of reading. If the conceptual object which is 'humanistic place', as argued by Tuan, is defined by sensory experience, then the manner in which a sensory experience of place is processed, read, captured, documented and remembered and *communicated* in the cultural record is an extension of the human experience. If I were to commit to words the array of sensory experiences I am receiving in the place where I am writing this, then those words could, in theory, be received by others, who have never themselves experienced this place, but the their sensory array would be very different from mine, even though I have 'caused' it by writing those words. The transience of humanistic place is subverted by the physical processes of writing, recording, printing and documenting (in much the same way that human movement is subverted by the fixity of photography, as described in Chapter 8). Place is, in other words, given form in text, and understanding that form is mutated through further readings and iterations through human cognitive processes.

Text as public space vs. private space

Viewing place as a construct which is created through the processes of experience and documentation highlights a parallel between text (and place) as private and public spaces. A text written by an individual person starts as a private document, akin to a private space or room with no other person present. It gains layers of exposure as (/if) it is read, copied and published – and if any of the content of the text involves a statement of, about, or relating to place, then this is similarly exposed and reinforced or challenged. The properties of access afforded by the distribution of text thus imbue it with some of the attributes of the 'public sphere' described by Habermas (Habermas, 1974). A place forming part of the 'public sphere', a town square for example, is not merely constituted of walls, buildings, a floor, monuments etc.; but also of the modes of ingress and egress, through-flow, rights of assembly, and the events that are (and are not) permitted there (and by whom). Put simply, a physical place has modes and models of access whose usage (according to Habermas) condition is of public significance, and so does a text. The assembly of private individuals in a public place, and the unfettered communication of the significance of that place, is the definitive part of what *makes* it a public space (ibid.); and unfettered public access to a text is the definitive part of what makes it a public asset.

There are various forms of public domain, or 'public sphere' in Habermas's terminology that text can occupy, and these differ with scale. The 'space' can be graffiti on wall, and thus localized to within a few feet; it can be a samizdat pamphlet, a book published by a mainstream publisher, and so on. Online, textual webpages, Tweets, blog posts etc., all have different reaches, scales and readerships, factors which contribute to, and by logical extension, expand and amplify, the 'emerging geographical events' of reading and writing (Hones, 2008). And text, whether in fact or in fiction, reaffirms and challenges socialized, collective and individual views of place in a variety of ways.

The post-Lefebvrian importance of place as a social entity created by communities which relate experiences through text, combined with the historical and archaeological importance of text as a framing medium of human culture, provides great potential for the use of digital methods in the understanding of place in human discourse. As noted above, textual communication and memory (and by extension the spatial properties of text) along with the polarizing dichotomy between author and reader, are a central component of human culture (Brown, 2001). It is no less a property of *digital* culture. The reversal of the reader-author polarization in the WWW age, with the advent of Web 2.0, User Generated Content and the 'prosumer', can be theorized in terms of the collapsing of place in relation to the process of reading. With Morettian approaches to critical quantification, GIS can be used to explicate the relationships between place references extracted from text, and the places themselves. Notable examples including the work of Ian Gregory on the literary geographies of the Lake District (Cooper & Gregory, 2011), in which the perspectives of different authors on the emotive properties of the landscape are extracted and critiqued using GIS.

In order to further this process of interdisciplinary engagement, it is necessary to consider the way in which spatial technology shapes textual communication, and the attendant processes of reading and understanding. The process of documenting place – a process which is always subjective – adds an interpretative layer which the digitally minded spatial humanist needs to unpick in order to understand the historical and/or authorial significance of a text's spatial properties. Taking a long view of the roots of spatial humanities in other discourses, it is possible to identify three categories of interpretative layer. While digital mediation has, as we will see, profoundly impacted on all of them, they represent sets of methods which can be applied in most literary-critical approaches to place in text.

First, there are the means by which we **communicate** place through text. In the digital world, this is done frictionlessly through the digital medium, abstracting place to standards such as WGS84. For example when a placemark is shared using Google Earth, this is done as a digital transaction, and the 'place' – or rather the abstractly digitized version of it – is renderable remotely, in exactly the same form as that in which it left the sender's machine (just as the text of an email appears in the recipient's inbox in exactly the same form as it leaves the sender's). Thus, as physical place is collapsed through the agency of electronic communication networks (see Chapter 2), so is the multivocality of described place rendered – and reduced – into a single 'voice' through the process of communication.

Second, there are spatially framed means used to **organize** text. The spatial organization of a text and the powerful associative functions of place combine to shape the transmission and reception of text through most genres, and in rhetorical technique. The incorporation of place in text aids the reader to retain and recall whole passages, and assists in the navigation (itself a spatially loaded term) of the text.

Finally, and most fundamentally, we have ways of **reading** text which foster the collective formation of place, in the manner described above. This is the most complex and discursive of the three layers, and underpins the transfer of a place recorded in text, and the spatialization of an idea in the reader's mind. There have been numerous studies on the impact of digital technologies on reading practices, especially the emergence in the 1990s and 2000s of digital reading devices, and how the digital device rapidly disrupted the practices of reading which evolved alongside ink and print technologies (Brown, 2001); but the aim here is to place these studies in the context of the broader questions of how readings of place in the digital age function.

These three categories – communication, organization and reading – describe the main properties of text-mediated human interaction, and they parallel the interactions afforded by physical space. Physical spaces also support processes of communicating and organizing, and are 'read', or received, or understood, by those in it, or in some way experiencing it directly. The digitization of text, the process of its shift from the written or printed page to the screen or device, represent the subverting of the text-bearing object – the book, the page, the inscription – by the process of – supposedly – seamless navigation of the text, unencumbered by the materiality of the object. In the same way, the collapsing of place around the process of reading enabled by the WWW subverts the materiality of space.

Communication

As noted in Chapter 2 and reiterated above, the Internet and WWW, from which most *narrative-forming* tools and practices used by the spatial humanities are ultimately derived, are first and foremost infrastructures for communicating information (including platial information) between machines, and users of machines. Therefore, related infrastructures for either creating text within them, or converting text into a form which they support (digitization), are also primarily machine-oriented communication infrastructures. This throws in to sharp relief the distinction between communicating place between humans and communicating place between computers. Communication between humans over distance can be spatially mediated by text; however, as described by Meyrowitz, the computerization of text impacts on the spatiality of human interaction itself. Meyrowitz argues that the ability to communicate simultaneously with multiple people using text, irrespective of physical distance, impacts on *what* humans communicate and *how* they communicate it.

As with other aspects of human(istic) space and place, these are functions of the medium of communication. The first structures of the World Wide Web of the early 1990s, which came three years after Meyrowitz was writing, contained, at their core, protocols for rendering 'hypertext', and allowing the user to browse between discrete units of hypertext in a nonlinear process which re-defined what it meant to read. At the same time, to this day, the WWW maintains the spatial vocabularies of the 'page' (a container of text) and the 'site' (a container of pages), whilst subverting and re-framing both. This shapes the cognitive formation of place through reading. Sheila Hones has argued (Hones, 2008), that the nexus between reading and writing itself is an 'always emerging geographic event', in which the context of the place in which both happen is a key conditioner of its reception. The WWW brings to the processes of reading and writing a frictionless instantaneity.

The primary purpose of digital texts and digital maps, and the technical infrastructures that underlie them, is to be easily connectible with one another, and with users. Online digital maps, such as that created on the WWW by Google Earth, Bing Maps, and other online mapping platforms, link to digital text through the shared semantic infrastructure of Extensible Markup Language (XML). This is also the format used in the development and deployment of many textual resources on the WWW, and has a key role in many Digital Humanities projects (Schmidt, 2012). The principle open geospatial standards used to encode place references, most notably the Keyhole Markup Language (KML) which underpins Google Earth and Google Maps, are also XML-based. This allows the user to navigate seamlessly between *connected* (/hyperlinked) online content. The seamless integration of spatial information, in the form of KML, with other forms of web content, was what drove the adoption of early web mapping platforms which accompanied the advent of Web 2.,0 in the 2000s (Graham, 2009), and it continues to condition the formation of space and place online today. But the limitless capacities for digital content made up of digital maps and digital text to integrate seamlessly comes at a cost of representational complexity and depth.

One response to this is to look at how humans engage with geospatial data. The affordances of neogeographic spatial content (see Chapter 5 for a full discussion) have been theorized as a means of expanding and enriching the connected GeoWeb (Goodchild, 2007). However, the methodological and epistemic implications of using the XML-based GeoWeb as a means of conveying textual place remain relatively little considered. By using geospatial standards such as KML and the WGS84 coordinate infrastructure of latitude and longitude and Mercator projection to georeference XML elements, spatial infrastructures in fact provide a profoundly un-social and un-Lefebvrian framework, which is completely arbitrary and abstract in the way in which it frames spatial relationships as schematic grids (Caquard, 2011). It is thus significant that online literary resources, especially those which make extensive use of Web mapping, often do not deal with mapping as a discursive practice in its own right, dealing instead with the ontologies of place and how different components of those ontologies are rendered in order to make them mountable on the Web, and visible to Web users (see especially Luchetta, 2017). This highlights the key between *reading* place and *communicating* it. The former creates a new set of spatial values, while the latter conveys the same spatial values from A to B, without creating any new interpretations. What is communicated is a set of binary data, a group of 1s and 0s which is reassembled at the remote computer and rendered in the remote browser in exactly the same order. Thus, a text whose spatial meaning is expressed in a Web map is automatically not subjected to any process of re-reading and post-representation, processes which the field of literary cartography has highlighted as being of primary concern in recent years. This tension between the frictionless communication of spatial information online, and the fuzzy, interpretative reading of place and in place in text, must be a core methodological concern for the spatial turn in the humanities.

A factor which deepens (or should deepen) this concern is the fact that the WWW (and digital media in general) treats text as a fundamentally binary and linear medium, whereas 'humanistic space' is – almost by definition – an analogue construct. As well as being one of the principle ways in which place, structured in the form of WGS84 coordinates embedded in XML elements, is represented online, XML provides a format for representing the *structure* of text quantitatively, a structure which, as argued above, is quasi-spatial. The P5 Text Encoding Initiative (TEI) Guidelines provide a set of structures and protocols for recording information about a place, its name, geographical location, and other attributes such as its population, political authority, and so on. It provides a hierarchical structure in which various kinds of platial/spatial information is attested within a text.

The TEI element <place> for example, denoting the word 'London' as a term with the semantic qualities of a place, can contain the sub-element <type> denoting it as a 'city', a figure for its population, and in what time periods, if applicable, different spellings or versions of its name were in use – 'Londinium', for example, being in use after AD 43 but before AD 1066, using the notBefore= and notAfter=. Such an approach is critical for building effective digital literary web map resources, of the kind reviewed by Luchetta (2017). Yet, as the TEI Guidelines website states, the purpose of this approach relates to

> [E]lements which may be used to record in a structured way data about places of any kind which might be named or referenced within a text. *Such data may be useful as a way of normalizing or standardizing references to particular places*, as the raw material for a gazetteer or similar reference document associated with a particular text or set of texts, or in conjunction with any form of geographical information system. (emphasis added) (TEI: Text Encoding Initiative, n.d.).

The purpose is to normalize and standardize the referents for the purpose of communication and the capacity to link seamlessly, while preserving the stylometric and semantic structure of the text itself. It is important to distinguish this from the purpose of representing or re-imagining how those places might be subjected to alternative understanding, re-readings or remediation in other contexts. This is not in any way to be construed as a criticism of the TEI's platial elements as a means of studying and analysing the humanistic geography of text, but rather to further accentuate the necessary distinction between textual place and reading textual place. It is evident that the emerging field of literary cartography is, essentially, attempting to do both. However any attempt to conduct *analysis* of place so communicated (as opposed to read) by itself will inevitably lead to criticisms that echo those of archaeological GIS of the 1990s and early 2000s: that it is overly positivist and reductionist, and that the essential meaning of the material – that gleaned through the reading of a text shared – is subsumed by that which can quantified and read by a computer. The value of digital representation of place therefore lies not so much in what it allows to be communicated, but in the ways it can *support* the process of reading.

Organization

References to place in text can likewise form the basis for creating structures of meaning. Moretti's idea of 'distant reading' was based partly on the idea that connections between concepts described in a piece of writing could be made by mapping the places referred to in texts and organizing them cartographically; and that so doing could inspire new insights into the text itself: 'Placing a literary phenomenon in its specific space – mapping it – is not the conclusion of geographical work; it's the beginning. After which begins in fact the most challenging part of the whole enterprise: one looks at the map, and thinks' (Moretti, 1998: 7). Implicit in this view is the idea that the analytic/narrative, quantitative/qualitative divide referred to above – also the distinction between the interpretative and experimental (Chorografia) and the measurable (Geografia) – can be transcended by the process of identifying and ordering points in the text which refer to place. The process of organization moves beyond the process of communication, as described above, to the ascription of meaning and context to the place reference, rather than just describing its structure. This idea of 'literary cartography as an interpretive practice' is picked up by Cooper, Donaldson and Murietta-Flores in their introduction to the recent volume, *Literary Mapping in the Digital Age:* '[T]he practice of literary cartography can be seen to take its orientation from a critical conviction in mapping as a practice that *enriches the reader's appreciation of the literary work of art*' (Cooper, Donaldson, & Murrieta-Flores, 2016: 7, emphasis added). The logical extension of this is that 'literary appreciation' through reading about a place leads to a form of spatial conception in the mind of the reader.

What links this view with the quantitative starting point of Moretti's distant reading is that place and spatial references provide a means of organizing and making sense of meaning in both individual texts and large corpora. Cognitive organization is distinct from the structuring/communicative approach of TEI, as it entails decisions that are not constrained only by equating the semantic meaning of a segment of text to a schematic tag.

We can use this distinction to identify parallels with the debates in archaeology discussed in the previous chapter. As highlighted there, one reason that archaeological data has received so much attention from GIS researchers and specialists, as opposed to other branches of the humanities, is that GIS, and digital mapping methods more generally, is a highly effective way to organize and analyse the material concerned. In most excavations and surveys, finds and the data relating to them are organized spatially, by trench, area, feature, or even site or region (Barker, 1993: 247–9; Lock, 2003: 164–83). A GIS plan of a site for example can show the layout of the trenches, the disposition of features such as walls and foundations, and of any finds recorded *in situ*. It documents an inherently spatial set of material in a manner that can be easily copied, published, analysed and distributed. Yet it cannot necessarily convey the excavator's (or anyone else's) spatial interpretation of that material.

A serious practical obstacle to deriving structured meaning from large quantities of text is that the texts available on a particular subject or genre will necessarily outstrip the capacity of any one human to read it all. According to one estimate, if one could read one book a day, it would take 40 lifetimes to read a library of one million books (Crane, 2006). The affordances of digital text for structure and search – which will be dealt with in greater depth in the next section – go some way to addressing the 'complexity deluge' facing the Digital Humanities (Anderson, Blanke, & Dunn, 2010), but still the tensions between communication/representation, and organization/understanding remain; as, most importantly, does the difficulty in determining what is significant and what is not.

Continuing the historical flavour of this survey of the spatial humanities, we cans see the organizational drivers of digital approaches to text in the earliest days of digital text itself. These problems of managing and organizing vast quantities of text became apparent in the period following World War II, when the quantity of research material available to scientists grew exponentially and the practical challenges for an international research community dependent upon a body of material growing at an unmanagable rate began to emerge. While masses of fictional writing and newsprint (which at that time was emerging as a dominant form of mass communication) had always been available, the growth of scientific writing at a rate which supposedly outstripped its potential to be utilized by the scientific community presented new challenges for reading, readership, and sense-making, and the construction of academic knowledge. In the context of the acceleration of technological development in the 1940s, the USA's wartime director of the Office of Scientific Research and Development, Vannevar Bush, wrote a classic article in 1945 entitled *As We May Think*. Here, he declared that '[t]here is a growing mountain of research … [b]ut there is increased evidence that we are being bogged down as specialization extends. The investigator is staggered by the findings and conclusions of thousands of other workers – conclusions which he cannot find time to grasp, much less to remember' (Bush, 1945: 112).

In this paper, Bush imagines the 'Memex' machine, a device which could transcend the 'artificiality of indexing' which relied on alphanumeric cataloguing sequences, placing information into mechanistic categories that were not expandable and thus unsustainable. The Memex would grow information structures organically as the volume of information increased, enabling a researcher to search, annotate, store and link microfiche documents, and explore the 'web of trails' that link the concepts they describe, whether or not those concepts are described by metadata. The idea was that a machine could algorithmically replicate the process of recalling ontological units of information by association, in the way that the human memory does naturally. Looking back on Bush's remarks from the second decade of the twenty-first century, there are some obvious conceptual and methodological parallels between what he imagined and academic Internet search engines, most notably Google Scholar, which ranks results according to their relevance to the user's search terms (Jacso, 2005: 1537–9).

Bush was writing in that period a little before the networking properties of the Internet, and subsequently the WWW described in Chapter 2, began to detach social and physical place. Before this however, place as attested in, and communicated through, text had particular cognitive properties of 'readerly' association, and the recall of information from text, centuries before Bush made his general point about the inadequacies of quantitative metadata for such recall processes, was a key part of rhetorical practice. This practice was used in Classical Greek rhetoric as a means information recall and delivery, a requirement in a world with limited written records and bureaucracy; and place was central point of reference; as it is to this day – the word *topos* (place/location) is, after all, also at the root of the English word 'topic'. Aristotle (Topics 163b 28–32) refers to this: 'For just as in the art of remembering, the mere mention of the places instantly makes us recall the things, so these will make us more apt at deductions through looking to these defined premises in order of enumeration' (quoted in the Stanford Encyclopaedia of Philosophy: https://stanford.library.sydney.edu.au/archives/win2011/entries/aristotle-rhetoric/#7.2).

Other writers linked extension of spatiality into concept recall, with the notion of *periegematikos*, or 'speech which leads one around' (Webb, 1999: 11), which both serves to increase the narrative impact for the audience, and acts as a rhetorical aid for the writer or speaker. In the former case, a text describes a place and, through telling, takes reader or listener on an *imagined* tour of it. This was an important element of Classical rhetorical training. It applies not just to place names, but to prepositions which identify particular places, whether actual or abstract. In his *Progymnasmata* for example, Aphthonius gives several examples of such rhetorical techniques, one of which is *Description*. This gives an idealized rhetorical example of how the student should portray, in words, a place. In his example, this is the citadel of Alexandria:

> A hill juts out of the ground, rising to a great height, and called an acropolis on both accounts, both because it is raised up on high and because it is placed in the high-point of the city. There are two roads to it, of dissimilar nature. One is a road, the other a way of access. The roads have different names according to their nature. […].
>
> After the steps is a gateway, shut in with grilled gates of moderate size. And four massive columns rise up, bringing four roads to one entrance. On the columns rises a building with many columns of moderate size in

> front, not of one colour, but they are fixed to the edifice as an ornament. The building's roof is domed, and round the dome is set a great image of the universe.
>
> (Heath, n.d.: Progymnasmata).

The use of prepositions to, after, round, implies movement through the (built) environment, and recreates it in the mind of the reader. This requires the scene, in the mind's eye, to be broken down into spatial components, which are then linked sequentially by the description. The mental visualization this creates is also a mental map: linked statements conveying geographical realities (the presence of a hill jutting out, its status as a high point, the dissimilarity of the two roads, the presence of steps etc.); which may be treated as ontological units within a mental map, which are given relational structure by the text. In much the same way, the text on a visual map owes its relational structure to the image – and to the values inherent in that image. It is therefore clear that rhetorical, visual and textual forms of spatial information are distinct, yet related, means of spatial representation.

The latter point, where spatial significance in text acts as a mnemonic for the reader/narrator, is highlighted by a digital mapping project of the University of Virginia's Scholar's Lab, which analysed the Catalogue of Ships described by Homer in Book II of the *Iliad*. The Catalogue documents the Achaean city states present at the battle of Troy, and enumerates where the ships came from. By visualizing the geolocations of the 190 place names in the chapter, and the narrative clusters in which they appear in Homer's text, the team was able to demonstrate that the arrangement of those clusters draws on a detailed knowledge of actual Greek geography, and they were grouped according to geographic logic, making them easier to recall by someone with such geographical knowledge (Evans & Jasnow, 2014: 324).

A similar project to apply digital mapping technology to an ancient author similarly shows the utility – and the interpretative limitations – of organizing and approaching text using place references. The HESTIA project, a study of Herodotus's *Histories*, produced a mappable database of place references extracted from the Herotodotean corpus in the Perseus Digital Library (Isaksen, Barker, Kansa, & Byrne, 2012).[1] A significant tension that this revealed was that geographical references in Herodotus are frequently ethonymic, i.e. they refer to the *people* of a place (e.g. 'The Athenians' rather to the place itself (Barker, Bouzarovski, Pelling, Isaksen, & Bouzarovski, 2010)). This distorts any attempt to represent the ancient textual geography of Herodotus according to modern cartographic standards and frameworks.

These cases show the significance of place and platial attestation as a means of organizing and recalling discursive streams of knowledge contained within text. The textual medium as a means of conveying meaning about place depends on its associative properties. When Herodotus writes of

1 http://www.perseus.tufts.edu/hopper

'Athens', he is referring to Athens in a sense which is conditioned by a number of factors and may, according to the associative properties of 'reading' place in a text, correspond to his reader's sense of what 'Athens' means. Platial attestations are meanings envisaged and realized by the author: in the languages of literary cartography, they all hinge on the fact that they represent an 'author's eye view' of the places described, and this may differ in its reception from the view of the reader. Especially as in the case of Herodotus, they reflect the authorial perceptions and conceptions of contemporary named sites and areas, which in turn draw on a range of social, cultural and literary sources.

At this point, we once again encounter the distinction (and tension) between chorographic place – which this process of cognitive recall represents, and the measurable, quantified place of the digital realm beyond the immediate field of observation. The discursive space recalled in the mind of the declaiming rhetorician can be contrasted to that 'recalled' by a computer. The same goes for the creation of spatial attributes from text. The identification of place names from (digitized) text, usually via automated processes, is *geoparsing*. In geoparsing, an algorithm identifies and tags words as place names, for extraction. The algorithms are prepared with 'training data', passages of text with the place names hand-coded to indicate the particular stylometric and semantic characteristics of place name words in that particular text (Caquard, 2011: 4). Following the process of extraction, the place names can be associated with coordinate references, the process of *georesolution*. Geoparsing and georesolution are basic processes in the construction of gazetteers from textual place references, a topic dealt with in more detail in Chapter 7.

As noted above, in the humanities much information is readily compatible with vectorized GIS models, so the construction of a (digital) gazetteer is often the first step in the preparation of humanities data for GIS work (Jessop, 2008: 40). More importantly however, place names associated with coordinate geometries have become a key part of the WWW and Semantic Web, with online gazetteers such as GeoNames.org playing a pivotal role in referencing places and allowing resources to be linked together and cross-searched based on common place references in them (Hahmann & Burghardt, 2010: 1) – a process of spatial recall which is simultaneously globalized, collectivized and mechanized. However, by default, the gazetteer model of representing place follows the hyper-empirical attribute > location model of GIS referred to above, and as with this model, it is not always appropriate for representing 'infological' discourse, (in Leszczynski's terms) or 'human' views of the world as portrayed in text and, by extension, of text itself (Ballatore, 2016: 8). This process, by which discursive text becomes structured data, is explored in more detail in Chapter 7.

Reading

The third and final layer, following the structuring and communicating of place in text, and the role of place in organization of text, is the process of reading. As Hones has pointed out, the geographies of reading shift across genres, and the *purpose* of the textual description is the most logical place to begin a discussion of 'reading' textual space. A place can be described fictionally or factually, and attested place names (which are themselves, in Western contexts, literary artefacts) can be embedded and referenced in text. The process of reading involves removing the place from its state of embeddedness and, very often, transferring it into another through onward communication.

The process of rendering place in text to place in another medium is exemplified by the descriptive passage of Aphthonius described above, and rhetorical interpretations of it: in this case, reading the text aloud remediates Alexandria as an entity frozen in the narrative immediacy of it being written down, from words on the page to spoken words. This has parallels with other rhetorical traditions, most notably *ekphrasis*, the process (and skill) of describing physical objects – in most cases art works – in words, of 'leading the eye around' in such a way as to re-create the object in textual form (Dunn, 2017). Whilst most scholars of ekphrasis are careful to distinguish between ancient and modern instances of the term, the earliest example of Classical ekphrasis is an example of mediation that text provides between the spatial imagination and other forms of expression. In other words, it illustrates the process of 'reading' place as a series of points where readers or audiences encounter text, and the place is made and remade at each point.

Ekphrasis as a means of 'seeing geography with words' manifests itself in the earliest literary example of the genre itself, the description of the Shield of Achilles in book 18 of the *Iliad* (see http://www.perseus.tufts.edu/hopper/text?doc=Perseus%3Atext%3A1999.01.0134%3Abook%3D18%3Acard%3D1, last accessed 25/7/2018). In this passage, Homer uses words to convey a narrative of themes from the epic canon depicted on the Shield, which was a metaphysical object which only ever 'existed' in abstract Homeric description. The earthly whole (*kosmos*) is represented in five concentric circles going around the edge of the Shield: from the middle, the earth and its constituent elements, marriages and feastings, the giving of 'righteous judgement', the besiegement of two cities, and scenes from rural life; all bound around the rim of the Shield by the River of Ocean. Brotton has drawn an explicit link between the Homeric description of the Shield as a device of ekphrasis, and as a map-like representation of the Earth, stating that 'Homer provides a *geo-graphy* – a graphic account of the earth' (Brotton, 2012: 4–9).

In his 'writing' of the Shield, Homer is describing a pre-Classical vision of the world's geography and humankind's place within it, with the human-under-the gods in the middle, the Ocean representing the furthest horizons of the physical world around the outside, and of human experience. Heffernan notes that by describing physical change in the landscape, such as the earth darkening behind the ploughmen, and the representational contrast between gold and tin, Homer is articulating a tension between narrative progression expressed in words (Heffernan, 1991) and the material fixity of the imagined 'object'. The stasis of the visual representation in the Shield's imagined material being contrasts rhetorically with the narrative flow of the events described; and this paradox is only possible in the medium of the text – and the encounters between the writer/rhetorician and their audience. One might thus develop Brotton's point, and note that any schematic map (and any text within it), is a fixed visual representation of land (and sea) at a particular time, which can either acknowledge or reject shifting human perceptions of that landscape. Phenomenological approaches to landscape have long recognized that landscapes are fluid, dynamic, changing (see Chapter 6), and conditioned by experience gleaned through habitation and movement. The written word, and readerly responses to place within it, is antithetical to the moving, temporal processes through which mountains are formed, rivers develop their courses, coast erosion changes the coastline and, of course, with the stories of human interaction with that landscape as it forms (Tilley, 1994; Barret & Io, 1999). In the place, and at the time of reading, a new – and static – 'version' of these places is formed in the mind of the reader. As Hones points out, the place and the time of reading frame and condition that (individualistic) act of place-making (Hones, 2008).

Returning to the themes introduced at the start of this chapter, the case of Achilles' Shield shows how reading place represented by the written or spoken word (i.e. whether through literacy or oral traditions) can stimulate the production of popular literary and artistic traditions. These also convey a particular (collective) spatial view of the past, which may or may not have any bearing on the real physical world (as, for example, with idealistically Classical view of space inherited from Ancient Greece, referred to above).

For example, one of the most powerful images of the post-Enlightenment period was that of the so-called 'golden age', a previous era when things were deemed to be 'better' than they are in the present. Invariably, golden age narratives are indelibly bound up with collective cultural perceptions of place – the narratives of collective memory described by Wertsch (2008). This manifests itself as nostalgic and anachronistic views of the physical world, and they are often associated with a notional rural and pastoral idealism, which was a common literary trope of the nineteenth century. In 1904 Martha Hale Shackford wrote 'The words *pastoral, idyllic, Arcadian* still move us; they surround us with an atmosphere of charm; they encourage those moments that are unstrenuous; they recall us to an early and almost outgrown freshness of feeling' (Shackford, 1904: 583). The aesthetic attachment to the rural idyll is often presented as a reaction to urbanization and modernization, and the shifting perceptions of the world that goes with them. These perspectives stress the manufactured and artificial nature of the rural idyll as an abstract place of the collective consciousness, separated from reality in time, space and concept (Bell, 2006). The descriptors 'nostalgic' and 'anachronistic' do not understate the importance of 'scenic aesthetics' in the reception and management of landscapes (Parsons & Daniel, 2002), or the power of rural imagery to influence both policy and perception (McLaughlin, 1986); but rather they acknowledge that the common 'golden age' view of past place is a product of cognitive separation mediated, very often, by text.

The perceived imperilment of the rural idyll by technology and (a pejorative view) of 'progress' – Massey's space-time compression – is often associated with traumatic events such as wars. These sharpen the spatially perceived distinction between the past and the present, readings which collectively form place amongst the readership(s) of text(s), and the distinction between reality and the ideal. These distinctions are especially evident in the writings of George Orwell (1903–1950), an author who traded heavily on spatial tension, with much of his political insight being expressed through his understanding of twentieth century audiences' preconceptions of generic locations with which they were familiar; and how these misalign with those conjured in his spatial narratives.

The key context for these writings was first the prospect, then the aftermath, of the World War II. Many remember Orwell for his novels of the immediate post-war period *Animal Farm* (1945) and *Nineteen Eighty-Four* (1949), two tales which revolve around the appropriation of familiar place(s) by despotic forces – the faceless and personless tower blocks of Oceania's London in the latter, their horror accentuated by narrative transposition to a dystopian 'London' of the future; and the microcosm of the eponymous (Manor) Farm in the former. In other Orwellian writings, this interplay between spatial expectation and spatial reality is more subtly studied. In *Coming Up For Air* (1939), for example, Orwell traces the story of 45-year-old George Bowling, who is trapped in suburban mediocrity defined by spatial uniformity. At the beginning of his narrative, George describes his environment in the present day (1939), at a time of impending war: 'Do you know the road I live in – Ellesmere Road, West Bletchley? Even if you don't, you know fifty others exactly like it' (page 9). This contrasts with the village of his childhood, Lower Binfield, which encapsulates the bygone age of 'before the war', with a rich array of imagery imparting a nostalgic view of the rural idyll. The book revolves around George's nostalgic reminiscences of the Lower Binfield of his past, its glaring contrast to his present, and his dismay and disappointment when he seeks to escape the drudgery of suburban life by rediscovering it – only to find that it has been overtaken by the march towards urbanization, and by the march towards the inevitable impending war. All that is left for him is to return, defeated, to the suburban spatial mundanity of Ellesmere Road.

Writerly approaches such as this, which appeal directly to collectivized readings and understandings of place, highlight the problem of applying digital methods to texts which describe 'lived in' space, even when that space is imagined. For example, a geodatabase containing all the place references in Arthur Conan Doyle's Sherlock Holmes stories could (e.g. in the course of an exercise of literary criticism of nineteenth-century crime fiction) contain lists of street names, named areas of London, named places outside London, and routes taken by characters between places. Each could be tagged with the attribute data of bibliographical references, literary context, and sentence context; and geocoded with the locational data of latitudes and longitudes for each entry. However, each process by which the information moves from Conan Doyle's text to the underlying database represents a process of reading followed by quantification and reduction, with judgements to be made about the geographical interpretation and precision of each geocoded data point, and its accuracy. Among the most problematic of the records would be the most famous place in the Holmes canon, the detective's rooms at 221B Baker Street. This is an address with a deep cultural attribute layer, in the form of dozens of references and descriptions in Conan Doyle's own writings, innumerable representations and re-imaginings in literature, film, radio, and popular culture, and a firmly established presence on the present-day London tourist scene. However, 221B Baker Street has not, and has never been, a real London address. Therefore, how would one geocode it as a record in a GIS database, following a process of reading? GIS is a powerful tool for the exploration of humanities material, but the reductionist nature of the link between attribute and location means it risks masking the reflexive and interpretative nature of the knowledge derived from that material.

These cases illustrate the point made above, that the discursive process of place-making through engaging with text (reading) must be distinguished from the purely quantitative method of structuring and representing text as it is written, following the stylometric forms laid out by the author or editor; and to the organization of spatial knowledge in text. These are all distinctions that have been driven by computational approaches and digital methods. All revolve around the *significance* of space and place in that process of engagement.

We should also consider the impact of digital media on the act of reading itself. It is important not to overlook the non-material structures which replace text-bearing material objects (such as books), and how these are still liable to guide reader behaviour. Brown, for example, notes that print-on-paper text provides 'tactile-topographical' reinforcement of the reading experience, and that readers 'spatially map the text [of print-on-paper] as they browse'. Importantly for this discussion he goes on to note that 'the reading process, and no longer the book as object, has become the central domain of digital texts' (Brown, 2001: 290–3). Exactly the same can be said of the 'place-making process' which text affords. The making of place through reference to abstract structures such as TEI, or the WGS84 grid system, serves a particular type of purpose: to locate places (or objects) on the Earth's surface accurately, or at least as accurately as possible. Spatial analytical affordances, including those offered by GIS (and used to good effect by the archaeological community) all operate through this lens of abstraction. Brown suggests that one of the key innovations being ushered in (when he was writing, in 2001) was full-text indexing, that this had the potential to revolutionize the reading process by taking the reader to any section of the text they were interested in by keyword. In the same way, the digital de-spatialization of text involves weighing the utilitarian requirement to locate an object accurately on the Earth's surface, versus the need to evince a critical understanding of what that place *means*, both inside and outside the context of the text itself.

Conclusion

Taking the methodological structures described above, the texts which most concern spatial humanities are, generally, those of formal knowledge and those of literature. As with archaeology, spatial computing approaches to both involve a spectrum between foci on mapping from positivist/empirical/objective viewpoints, to a reflexive view of how (textual) space works in human terms, which draws more heavily on the theory of history and literary criticism. However, across this spectrum, the shift from application to theory identified by Gregory and Geddes can be discerned. The introduction of GIS methods, especially vector GIS, into the fields of history and literary studies problematizes the distinction between the two, with computational methods which, while sharing many technological characteristics, have facilitated the formation of two very distinct new fields.

There is much, however, that the fields of literary cartography and spatial archaeology can learn from each other – if one develops the parallel described above between physical and textual spaces, then methodological parallels emerge between rendering, visualizing and mapping the physical spaces of interest to archaeologists, and rendering, visualizing and mapping the textual spaces of interest to literary cartographers.

A key aim which literary cartography shares with the spatial humanities is the development of an understanding of the spatial significance of text which goes beyond the simply representational (Rossetto, 2016). However the 'simply representational' is precisely what the spatial substructures of the WWW, TEI and WGS84 are designed and purposed to deliver. What – the spatial humanist might ask – is the potential, or even the purpose, of digital mapping for literary cartography? It seems that the answer to this lies, as Moretti suggested, in the ability of these quantitative structures to transcend the limitations of what a single human is able to read, much as Bush foresaw with his notional Mimex machine, and to think creatively about both the discursive and the structural aspects of text. Potential new understandings of the spatiality of text arise from the ability to aggregate, amass and conjunct vast quantities of material, not to read a million books (as Crane muses), but to be able to extract spatial significance, and spatial meaning, from a million books.

This chapter argues that there are three components to this: communicating, organizing and reading. The cases surveyed briefly above show that these are closely interconnected, and that digital methods form a component of each, but the limitations must be recognized. XML and KML are protocols rooted in communication, and their purpose is to provide quantifiable structure; they can form *part* of organizing the places (actual and abstract) in a text, although such modes of organization go back at least to Ancient Greece. Finally, those places are received and read by audiences and readerships. The logical corollary to this is that place-formation from (/by) text is an ongoing and nonlinear process. It is a networked, social and communal process. Undertaking this process in a detached and critical manner imbues it with the attributes of what is defined in Chapter 1 as 'deep mapping'.

References

Anderson, S., Blanke, T., & Dunn, S. (2010). Methodological commons: Arts and humanities e-science fundamentals. *Philosophical Transactions of the Royal Society A: Mathematical, Physical and Engineering Sciences*, 368(1925).

Andrews, J. H. (2009). *Maps in those days: cartographic methods before 1850*. Four Courts Press.

Ballatore, A. (2016). Prolegomena for an ontology of place In H. Onsrud & W. Kuhn (eds.), *Advancing geographic information science*, 91–103. Needham, MA: GSDI Association Press.

Barker, E., Bouzarovski, S.,, Pelling, C., Isaksen, L., & Bouzarovski, S. (2010). Mapping an ancient historian in a digital age: the Herodotus Encoded Space-Text-Image Archive (HESTIA) How to cite: Mapping an ancient historian in a digital age: the Herodotus Encoded Space-Text-Image Archive (HESTIA). *Leeds International Classical Studies*, 9(1).

Barker, P. (1993). *Techniques of archaeological excavation*. Psychology Press.

Barrett, J. C. & Ko, I. (2009). A phenomenology of landscape: A crisis in British landscape archaeology? *Journal of Social Archaeology*, 9(3), 275–294. Available online: https://doi.org/10.1177/1469605309338422

Bell, D. (2006). Variations on the rural idyll. *Handbook of Rural Studies*, 149–160.

Bolter, J. D. (1991). *Writing space: The computer, hypertext, and the history of writing*. Lawrence Erlbaum Associates, Inc.

Brotton, J. (2012). *A History of the world in twelve maps*. London: Allen Lane.

Brown, G. J. (2001). Beyond print: reading digitally. *Library Hi Tech*, 19(4), 390–399.

Bush, V. (1945). As we may think. *Atlantic Monthly* (July 1945), 112–124.

Caquard, S. (2011). Cartography I: Mapping narrative cartography. *Progress in Human Geography*, 37(1), 135–144

Cooper, D., Donaldson, C., & Murrieta-Flores, P. (eds.). (2016). *Literary mapping in the digital age*. Abingdon: Routledge.

Cooper, D. & Gregory, I. N. (2011). Mapping the English Lake District: A literary GIS. *Transactions of the Institute of British Geographers*, 36,89–108.

Crane, G. (2006). What do you do with a million books? *D-Lib Magazine*, 12(3).

Dunn, S. (2017). Praxes of 'The Human' and 'The Digital': spatial humanities and the digitization of place. *GeoHumanities*, 3(1), 88–107.

Earley-Spadoni, T. (2017). Spatial History, deep mapping and digital storytelling: archaeology's future imagined through an engagement with the Digital Humanities. *Journal of Archaeological Science*, 84,95–102.

Evans, C. & Jasnow, B. (2014). Mapping Homer's Catalogue of Ships. *Literary and Linguistic Computing*, 29(3), 317–325.

Ferdinand, P. (2000). The internet, democracy and democratization. *Democratization*, 7(1), 1–17.

Goodchild, M. F. (2007). Editorial: Citizens as voluntary sensors: spatial data infrastructure in the world of Web 2.0. *International Journal*, 2,24–32.

Graham, M. (2009). Neogeography and the palimpsests of place: Web 2.0 and the construction of a virtual earth. *Tijdschrift Voor Economische En Sociale Geografie*, 101(4), 422–436.

Gregory, I. G. & Cooper, D. (2010). GIS, texts, and images: New approaches. *Poetess Archive Journal*, 2(1), 1–22.

Gregory-Guider, C. C. (2004). 'Deep Maps': William Least Heat-Moon's Psychogeographic Cartographies. *eSharp, Journeys* 4,1–17.

Habermas, J. (tr. Lennox, S., & Lennox, F). (1974). The public sphere: an encyclopedia article (1964). *New German Critique*, 3, 49.

Hahmann, S. & Burghardt, D. (2010). Connecting linked geodata and geonames in the spatial semantic web. *Proc. of GIScience 2010 Extended Abstracts, Purves, R. and Weibel, R. (eds.),* (1999). Available online: http://www.giscience2010.org/pdfs/paper_165.pdf

Heath, M. (n.d.). Aphthonius Progymnasmata. Available online: https://people.umass.edu/dfleming/E388 Aphthonius Progymnasmata.pdf. (retrieved 25 July 2018).

Hodder, I. & Orton, C. (1976)*Spatial analysis in archaeology.* Cambridge: Cambridge University Press.

Hones, S. (2008). Text as it happens: Literary geography. *Geography Compass*, *2*(5), 1301–1317.

Isaksen, L., Barker, E., Kansa, E. C., & Byrne, K. (2012). GAP: A NeoGeo approach to classical resources. *Leonardo*, 45(1), 82–83.

Jacso, P. (2005). As we may search – Comparison of major features of the Web of Science, Scopus, and Google Scholar citation-based and citation-enhanced databases. *Current Science*, 89(9), 1537–1547.

Jessop, M. (2008). The inhibition of geographical information in digital humanities scholarship. *Literary and Linguistic Computing*, 23(1), 39–50. Available online: https://doi.org/10.1093/llc/fqm041

Lock, G. R. (2003). *Using computers in archaeology: towards virtual pasts.* Psychology Press.

Luchetta, S. (2017). Exploring the literary map: An analytical review of online literary mapping projects. *Geography Compass*, 11(1), e12303.

McGann, J. (2006). From text to work: Digital tools and the emergence of the social text. *Romanticism on the Net*, 16(41–42),49–62.

Mclaughlin, B. (1986). Rural policy in the 1980s : the revival of the rural idyll. *Journal of Rural Studies,* 286(2), 81–90.

Moretti, F. (1998). *Atlas of the European novel, 1800–1900*. London: Verso.

Olwig, K. R. (2011). Choros, chora and the question of landscape. In S. Daniels, D. DeLyser, N. J. Entrikin, & D. Richardson (eds.), *Envisoning landscapes, making worlds: Geography and the humanities* (pp. 44–54). Abingdon, UK: Routledge.

Parsons, R. & Daniel, T. C. (2002). Good looking: in defense of scenic landscape aesthetics. *Landscape and Urban Planning*, 60,43–56.

Rossetto, T. (2013). Theorizing maps with literature. *Progress in Human Geography*, 38(4), 513–530.

Rossetto, T. (2016). Semantic ruminations on 'post-representational cartography', *International Journal of Cartography*, 9333(April),1–17.

Saunders, A. (2016). The spatial practices of writing. In D. Cooper, C. Donaldson, & P. Murrietta-Flores (eds.), *Literary mapping in the digital age* (pp. 147–160). Abingdon, UK: Routledge.

Schmidt, D. (2012). The role of markup in the Digital Humanities. *Historical Social Research / Historische Sozialforschung*, 37(3), 125–146.

Shackford, M. H. (1904). A definition of the pastoral idyll author. *PMLA*, 19(4), 583–592.

Sui, D. & DeLyser, D. (2012). Crossing the qualitative-quantitative chasm I: Hybrid geographies, the spatial turn, and volunteered geographic information (VGI). *Progress in Human Geography*, 36(1), 111–124.

TEI: Text Encoding Initiative. (n.d.). Available online: http://www.tei-c.org,(retrieved 15 July 2018).

Tilley, C. (1994)*A phenomenology of landscape: Places, paths, and monuments.* Oxford, UK: Berg.

Webb, R. (1999). Ekphrasis ancient and modern: The invention of a genre. *Word & Image, 15*(1), 7–18.

Wertsch, J. V. (2008). The narrative organization of collective memory. *Ethos, 36*(1), 120–135.

5 Spatial humanities and neogeography

Introduction: user generated place

Much of today's global human communication is conducted through the Internet and, in this process, vast quantities of spatial data are created on an annual, daily and hourly basis. This is done through the tools and platforms of the GeoWeb, whose history as a subset of the WWW was outlined in Chapter 2, and whose relationship to the organization of text was explored in the previous chapter. The GeoWeb is a loose federation of standards, content, users, services and data online which share a primary concern with mapping. It is differentiated from the Internet and WWW more broadly by its purpose, which is to aggregate and expose data with a geographical component by 'integrating cartographic geodata with geotagged hypermedia' (Scharl, 2007: 3). Offen defines it as 'the combination of new technologies and applications that seek to georeference information on the web for location searching' (Offen, 2012). Lake and Farley on the other hand stress the aspects of the GeoWeb which support integration and discovery: '[T]he Geospatial Web is an integrated, discoverable collection of geographically related Web services and data that spans multiple jurisdictions and geographic regions … the global collection of general services and data that support the use of geographic data in a range of domain applications'. The GeoWeb is therefore a disparate and loose concept, but the popularity of its services, products and applications, and its centrality to daily life, has put it at the heart of debates around spatial literacy, and the means by which casual users, students and researchers engage with place (Bednarz & Kemp, 2011, see also Chapter 1). As will be seen below, it is also a key tool for spatial research in the humanities.

The role of the GeoWeb as an agent of spatial literacy and a tool(set) of the spatial humanities is complicated by its human and technical history. Like the Internet and WWW of which it is part, the GeoWeb has no single designer, or even a single set of design principles. It is partly a product of the development of mobile technology which enables users to contribute data to it as 'voluntary sensors' (Goodchild, 2007) as well as consuming it; of GPS systems which allow the data thus created to be georeferenced in real time (Junglas & Watson, 2013); of the communication infrastructures that give the real-time mobile device coverage that enables georeferencing (Hill, 2009); the applications that enable users to use spatial data, and the standards which underpin those applications. But most importantly of all, it is defined by the very visible presence of a set of large, monolithic web mapping platforms such as OpenStreetMap (OSM), Google Earth, Bing Maps and Yahoo Maps, which emerged as the use of geospatial media gained wide uptake and usage. These, more than anything, embody the public face of what the GeoWeb is today, and have come to guide large swathes of WWW usage, sometimes obviously, sometimes not. These platforms also frame numerous debates between the merits of Open Source software (like OSM), and proprietary applications (like Google Maps). Inevitably such debates become tied up with wider questions of internet freedom and neutrality, and the commercial and cultural power of multinational corporations.

The flow of crowdsourced geographic data accompanying the development of spatial Web 2.0 platforms has, in recent years, grown from a torrent to an 'exaflood', to use the term of Sui (Daniel Sui, Goodchild, & Elwood, 2013) Tens of millions of geographic data points are captured each day by both mobile and desktop hardware devices, representing billions of individual perspectives and experiences, however fleeting and ephemeral those experiences might be. The rhetoric of Web 2.0 in the second half of the 2000s stressed this as a form of unconstrained multi-authorship, of democratization, but this has been questioned, as is explored below. Some (e.g. Haklay, 2013) question instrumentalist notions of 'open' mapping platforms such as OSM, which separate means from ends, and tools from applications. The instrumentalist, or Marxist/Darwinist, view of technology assumes it is a value-free medium which is inherently devoid of any political or social agenda, and is therefore separate from human progress and the development of social complexity. Haklay identifies four categories of use to illustrate engagement with these technologies: 'Meaning hacking', ''Use', 'Shallow technical' and 'Deep technical', which correspond broadly to a conception of 'lightweight' contribution in terms of meaning hacking (Haklay, 2013: 66). This holds that the online map can be manipulated by low-effort interventions, and is the opposite of 'heavyweight' interventions at the 'Deep technical' end, which require highly specialized technical knowledge, available only to a small elite. Such contributions are made according to the motivations and rewards of community recognition and the advancement of personal standing in that community.

Neogeography platforms thus depend on very different levels of engagement, since those engaging inevitably bring their own perspectives and their own interpretations to this work. Therefore, the question is not so much whether platforms like OSM are value-free, rather it is what kinds of value they generate. One way of viewing this is through the demographics of those participating. There are well-documented disparities in the user bases of neogeography, especially in gender. In Budhathoki & Haythornthwaite (2013)'s survey of users for example, 96.2 per cent of respondents were male. This further calls in question the claim that VGI and geographic Web 2.0 tools have a democratizing effect. This also inevitably conditions the kinds of narrative that can be formed through the volunteered creation of geodata.

The importance of the GeoWeb as a distinct part of the online world is the result of a number of user-oriented factors. This reflects the definition of the GeoWeb made above (Chapter 2) as a subset of the WWW, whose services and data infrastructures are exclusively concerned with mapping. For example, maps take on particular significance for individual decision making if they are used in specific places, e.g. if they are being used to locate particular services or places. This significance can be further enhanced and nuanced when the detail and coverage of a map is enriched by annotation of the area and/or content that the map represents with *other* services or locations. These might gain visibility to users coming to the area for the first time through being annotated by other users, or through means mediated algorithmically, such as search engine results (Ford & Graham, 2016). Neogeographic mapmaking is a continuously recorded conversation about place, among a network of people (who may or may never meet in the physical world). The importance and utility of a digital map used *in situ* is thus – potentially – context-specific and bidirectional. Given this conversational nature of online mapmaking (however mediated), it is significant that the loose federation of technologies and practices which the GeoWeb represents emerged in the mid-2000s, at the time when the Web 2.0 movement was redefining the WWW as a bidirectional platform, where users could produce and publish data as well as consume it (Turner, 2006). The most significant characteristic that the GeoWeb retains to this day is the Web 2.0-like capacity it affords for users to *create* as well as *consume* map content (Haklay, Singleton, & Parker, 2008).

So what of the GeoWeb's role in scholarship? The spatial humanities is a critically driven field with a greater emphasis on the past than on the present, and – as noted in Chapter 4 – the GeoWeb is primarily a network of hyper-representational maps which are built, and exist, in order to communicate information, not to map it. Topics related to the GeoWeb have, however, received some attention more recently from those concerned with the humanities, usually in the context of building web resources for humanities scholars (e.g. Luchetta, 2017). However, much of the critical theory around the GeoWeb and its associated practices, has come from the domain of human geography, especially those branches of human geography associated with Critical GIS studies and, more especially, Public Participatory GIS (PPGIS) (Haklay, 2013). A key question for the spatial humanities therefore is what can those domains which study place in human discourse through the critical lenses of the humanities (see Chapter 1) learn from those which study place in the *contemporary* discourses of human culture, seen through the lenses of the GeoWeb?

Critical approaches to the contemporary GeoWeb

To begin with, most human geographers who have tackled the subject acknowledge the separation of practices and technology, maintaining a distinction between the *technical infrastructures* of the GeoWeb and the *social and cultural processes* it enables (Zook & Graham, 2007b). These cultural processes, and their consequences, have spawned numerous terminological variations for the unconstrained creation of geodata by the general public. These include 'neogeography' itself (although this is not, itself a new term), Volunteered Geographic Information (VGI), 'ubiquitous cartography', and 'web mapping', all of which are enabled by the federated, disarticulate constellation of GeoWeb standards and platforms referred to above. Most importantly, this range of terms reflects the wide range of uses to which maps based on distributed collective data are put (See et al., 2016).

The central feature common to all definitions and nomenclatures of GeoWeb practice is the – supposed – open, borderless and frictionless manner in which it allows geodata to be created by the public, producing maps that are the result of collective amateur effort, rather than of professional cartographers or geographers. This is seen as being both culturally and practically opposed to the process of producing maps through the official agencies of corporations or governments, such as the Ordnance Survey in the UK, or the US Geological Survey, and supra-governmental organizations like the UN or the European Union. Goodchild stresses this aspect in his classic 2007 framing of VGI, reflecting the rhetoric of the interactive web that was opening up at this time:

> [Mapping agencies] represent the top-down, authoritarian, centrist paradigm that has existed for centuries, in which professional experts produce, dissemination is radial, and amateurs consume. … The world of VGI is chaotic, with little in the way of formal structures. Information is constantly being created and cross-referenced, and flows in all directions, since producers and consumers are no longer distinguishable. (Goodchild, 2007: 29)

For human geographers concerned with the human dynamics of the production of geographic knowledge, this deconstruction of the producer/consumer relationship, unconstrained by formal assemblages of power or accountability, is the most significant consequence of the growth of GeoWeb technology. After all, the ability to record a feature's geometry and location on the Earth's surface using a pin or pins on Google Maps bypasses a large panoply of skills, experience and methodology which the same process would involve in a professionally produced map. So what key difference to the outcome/product/asset produced does the amateur or collective process of VGI/neogeography make?

The main difference is that the product will, inevitably, reflect the collective biases and partiality of the user base, and how the technical infrastructure channels and represents those biases. The definition which most explicitly addresses the dichotomy between services/infrastructure and the data (imbued with cultural and social bias) is Zook and Graham's description of what they call '*DigiPlace*'. In their discussion of DigiPlace, Zook and Graham argue that 'cultural imperatives' which drive the creation of geodata on GeoWeb platforms such as Google Maps are conditioned by the algorithms that produce it. However, those imperatives are obscured by the flat, representational nature of the maps themselves (Zook & Graham, 2007a: 480).

Zook and Graham also point out a key way in which the WWW and the GeoWeb overlap. The visibility and prominence of individual 'DigiPlaces' on mapping platforms are determined not by social or cultural values, or how important the place is to any particular community, but rather by ranking algorithms employed by major search engines such as Google (Zook & Graham, 2007b). This represents a commodification of relative visibility and networks of value, stemming entirely from internet search behaviour and the flow of advertising revenue, rather than any cartographic principles. Arguably therefore, it contradicts the implied notion that a platform such as Google Maps provides neutral or unbiased mapping; or that it presents a carto-representation which formally embeds parity between places in a way that, say, an 'authoritarian, top-down' mapping agency of the kind described by Goodchild might (or at least might aspire to). It also undermines the notion, critiqued by other human geographers, that VGI/neogeography – or whatever one calls it – represents a democratizing and empowering influence for voices which might otherwise be marginalized (see below).

The GeoWeb and research: Citizen (Spatial) Humanists?

This brings us back to the question of what critical and/or academic uses the GeoWeb might be put to in the spatial humanities. If 'DigiPlace', or the products of neogeography, or VGI, or whatever one calls it, is unstable, why should those engaged with the critical investigation of humanities research questions – which, as noted above, extended into the past rather than the present – have any interest in it? Or why, to put it subtly differently, is the GeoWeb, and the data and processes it enables – which, as described above, are fraught with cultural and social bias – be important to the spatial humanities?

To approach such a question properly, it is necessary to re-state the distinction between GIS, which is a tool of analysis and calculation, and the GeoWeb, which is derived from the WWW's toolsets of communication. There are clear cases of use, practice and theory of the of the former, as applied to humanities research questions (even though – as discussed in previous chapters – key thinkers of the field are now reducing the emphasis on application-oriented cases of GIS use in the humanities in favour of the methodological commonalities which link them). It is less clear, in the context of previous debates, what is gained for the epistemologies of the humanities by drawing on (communicative) GeoWeb resources, when the main characteristic of the latter is that they have been created through the modes of mass production enabled by platforms such as Google Maps, or indeed any social medium which carries the ability to georeference data according to the standards which the GeoWeb understand. The relationships between these new methods of doing geography – *neo*geography – and humanities methods are less obvious than those of GIS, which are rooted in analytical methods (and motivations) which are altogether more familiar to humanists.

The answer, explored in more detail below, lies principally in the emerging depth and complexity of the relationship between 'citizen humanists', people who do not access, create and use geodata through institutionally enabled platforms, but rather are users of the (Geo)web outside academia and other corporate entities, who create content and contribute to research in the context of academic crowdsourcing (see Hedges & Dunn, 2017). The 'expert vs. non-expert' paradigm has been discussed elsewhere. Offen reiterates the importance of user-generated content in his definition of neogeography as 'non-expert, user-generated geographical content for collective action' (Offen, 2012: 567). Implicitly therefore, neogeography is a suite of practices which should reinforce the distinction between those researchers and general internet users. But does it have to? The key question is in what ways are the new forms of connectivity linking expert versus non-expert activity on the GeoWeb?

In the context of the spatial humanities, the act of asking this question highlights one of the special characteristics of spatial research and spatial content. In most other areas of the (digital) humanities, the realms of online research easily and unproblematically reflect the established institutional boundaries of the offline world. For example, TEI, the method of structuring text discussed in the previous chapter, has not been subject to suggestions that it has blurred the distinction between professional text scholarship and, say, members of the contemporary public creating and reading text online in the blogosphere, or on social media. Furthermore, most intelligent readers or users would not have any trouble distinguishing between a blog post, and a section of manuscript text, such as (for example) the Perseus Digital Library, which has been subject to expert TEI intervention. Nor are there suggestions that the digitization of text and its dissemination online has created risks of appropriation by large corporations or other interests, in the way that has been argued for neogeography (see below). There is a mass of critical method, theory and tools underpinning textual approaches, such as those discussed in the previous chapter, which buffers it from generic text creation on the Internet – but the creation of geographic data, which is based on first-hand knowledge, experience or expertise, is not such a skill.

Motivations

The huge quantity of voluntarily created geodata on the WWW begs the question of why people outside institutions create it in the first place. Users contribute text to the WWW through blogs, tweets, social media etc. for a vast array of reasons, but the reasons for contributing to mapping platforms are more constrained (and, perhaps, better researched). There are many forms in which GeoWeb data is 'produced' (of which more below); but in general, the most common method of production is 'pinning' (another physical object and process which has found its way into the geo-digital lexicon) geographic features to the major mapping platforms. The sheer scale of data created by open public contributions to these platforms indicates a vast reservoir of motivation in some form: in 2016 for example, *TechCrunch* ran the headline 'Google Maps is turning its over a billion users into editors' (https://techcrunch.com/2016/07/21/google-maps-is-turning-its-over-a-billion-users-into-editors, accessed 25 July 2018). Understanding what motivates over a billion individuals to participate in a mapping platform, and thus to contribute data voluntarily, is key to understanding the value, biases and possible uses of the data itself.

Collective reasons for creating geography are easier to understand than individual ones. In Chapter 2, I sought to highlight the principal motivations (at the same time rationalist and nationalist) of the early chorographers at the beginnings of modern scientific cartography. It can be shown in these cases that these chorographer/cartographers had a range of reasons for embarking on their labourious journeys – nationalistic, religious, and intellectual – but a feature of most of their texts is that a clear rationale is given at the start for *why doing so is important*. It is therefore relatively easy to discern why, in individual cases, the chorography was written, the maps created, the places described. Even more, the motivations that national mapping agencies have in the production of geodata are transparent, and made plain by the organizations concerned. They serve particular customers, whether in the commercial market, or governments needing to provide services; they have budgets, and will usually have some form of mission statement or other articulation of purpose (Perkins, 2007; Goodchild, 2007: 211–12). VGI, or neogeography platforms may state their own reasons – OpenStreetMap, for example, states that it 'emphasizes local knowledge', and that its community is 'diverse, passionate, and growing every day' (see https://www.openstreetmap.org/about). Google MyMaps says it is there to '[e]asily create custom maps with the places that matter to you.... View custom maps in the Google Maps mobile app so you can find places on-the-go.... Find all your maps in the Your Places menu on Google Maps.' But again, these are the motivations of incorporated entities, not the individuals. In the world of neogeography, in other words, organizations need to explain their motivations for mapmaking in a way that individuals do not.

Why would a billon private citizens with little or no formal training in mapmaking, want to become 'editors' of Google Maps? What drives these individuals to share the data in the first place? Those who contribute to mapping platforms have disparate, complex reasons for doing so, but in the so-called 'long tail', these reasons are very often *intrinsic*, i.e. the person wishes to answer a question, or solve a problem, or record experiences that are specific to their individual requirements at a unique moment. They are driven by needs that are particular to the individual. Zook and Graham suggest that a major motivation for participating in the GeoWeb as a data producer is that one wants to search for particular locations/services or objects, and the data is captured in the course of this activity. It follows that generic geodata about individual subjects or objects – with their main attribute being their locations, referenced to WGS84 web-readable standards – will be created in proportion to the level of interest in those objects which is shared by individual (and uncommunicating) GeoWeb users. In turn, it follows that visibility on GeoWeb platforms is tied to the same factors that cause particular results to appear higher up in Web search rankings. Successful leveraging of this passive/consumerist model is behind the success of the mapping arms of corporations such as Google. This stands in contrast to the motivations of cartographic organizations, or agencies, who produce maps professionally for market consumption, in response to consumer need, and in the expectation or hope of reward (financial or social etc) for doing so.

There are many analyses of data derived collectively from GeoWeb users, and comparisons of the accuracy (or otherwise), epistemology and infrastructures of these data and those produced by mapping agencies (Haklay, 2010); yet the distinction between professional and non-professional cartography is less often expressed in terms of the motivation(s) of those creating the data. Yet a key way to express the amorphous distinction between the GeoWeb and the broader Internet is in terms of the types of things people use the GeoWeb for compared to those for using the broader WWW. Some elements of the global digital infrastructure, such as KML, GPS devices and web mapping platforms, are *solely* concerned with aggregating data from many users to create online maps and geodata. The WWW, on the other hand, was developed as a means for storing and linking documents containing text. In contrast, the GeoWeb's infrastructures emerged in response to the more specific and practical need users have to locate objects and places on the Earth's surface and, certainly in more recent times, to way-find. This is a very specific type of motivation, and it helps to explain the great growth of the GeoWeb in response to allied technological developments such as GPS. Put simply, as GPS became available to more and more people via cheaper and easier to use devices, so the potential to use and create geodata grew, driven by this specific class of motivation. This trend in the use and production of global geodata has been harnessed by corporations making use of Location Based Services (LBS), and is highlighted by the importance that LBS have taken in contemporary digital culture and commerce, and the significance of geo-location in push-pull economics (Jiang & Yao, 2006).

Scale: the paradox of neogeography

Much collective GeoWeb data is created through observation and recording, the aggregated digitized outcomes of billions of individual perspectives. Such data is often created using hand-held devices, which provide an inherently personal user experience at the scale of the individual. Spatial UGC created in real time in this way (which is not all spatial UGC, as explored below) is thus paradoxical. When data is recorded and georeferenced on a smartphone, it is captured at an individual – perhaps chorographic – scale; yet when shared it becomes an indivisible part of a map of exponentially larger – perhaps geographic – scale. Take, for example, a restaurant review on a site such as TripAdvisor. At the core of the review is a geolocation defined by a point, the restaurant itself. This is inherently 'geographic'. The reviewer's opinion is quantified by giving a grade out of five for various aspects of the experience. The body of the review itself however is a narrative describing the author's experience in a purely subjective manner. The quantified and measurable aspects of the review at large scale – the rating, and the geolocation, are disposed to being aggregated into an articulated-seeming mass of thousands of other reviews, giving a composite overview of the sub-regional or regional picture at *smaller* scale – the numbers of a particular type of restaurant in a part of a city; the numbers of highly rated restaurants in the same area, and so on. It is this aggregated data which guides searching and retrieval behaviours of those exploring the dataset; and of course, the methods of exploration which are available to the user are determined by the design of the app. In a sense therefore, the GeoWeb collapses the distinction between geographic, or measured, place and chorographic, or described place which, as the previous three chapters argue, has always informed the discourses of human geographical experience.

Even where neogeographic data creation does not rely on direct, georeferenced documentation of experience (such as a restaurant review), but rather on applying geographic properties to other types of data, such as georeferencing photographs, the paradox of scale persists. Contributors to open platforms, such as OpenStreetMap, which georeferences infrastructures such as roads, railways, physical features, buildings and so on, either from existing paper maps or from satellite photographs, are still using and documenting their own knowledge of particular points of the Earth's surface at a de-scaled level, creating and transcribing maps by transferring data from one medium to another. According to See et al., only a small proportion of GeoWeb users engage in this type of activity (12 out of a sample of 100 – (See et al. 2016: 11)). But there is a key difference between this, and those who contribute from the personal (and thus subjective, or quasi-subjective) scale: the former *seek* to conform to empirical cartographic guidelines, to include geometrically correct points and exclude (or correct) incorrect ones; to construct accurate spatial footprinting (= Goodchild's 'formal structures'). The geometry is the most important element of the contribution. The latter however is a subjective viewpoint, which requires only a correct point associated with a piece of information, such as a spatial value – for example a restaurant review, a GPS trace, transportation data and so on. Most significantly, it is the latter form of geodata that most refer to when discussing the cultural impact of the GeoWeb, the practices versus the technology, and the biases it encodes.

Freed from the entrenched distinction between geographic and chorographic place, neogeographic geodata is prone to manipulation. We must be wary of any suggestion that the individualization of data which ostensibly underpins crowd-contributed geographic information makes it inherently open, or that it necessarily challenges or subverts existing commercial, social or political structures. Neogeography is certainly not subversive: it must be distinguished from 'counter-cartography' movements, such as the Paris Situationists, which actively resisted the cartographies of power and the authorities they represented (see Chapter 2). Such movements emphasize the perspective, and the needs and motivations, of the individual, and individual's embodied experience of the city, over the perspective of the authoritative municipal mapping agency. Critically however, they have their own defined ideological and political drivers. GeoWeb platforms place the means of cartographic production in the hands of individuals, allowing them to contribute their individualized perspective to a bigger whole, and by georeferencing 'chorographic' observations. By doing so, they subvert the 'formal structures' of Goodchild. In both cases – observed contributions, and transcribed contributions – the formal, authorial and ideological structures of cartography are replaced by other kinds of structure, which transcend those to which Goodchild was referring – especially structures of scale. However, because the formal, web-standardized grammars and vocabularies of the GeoWeb apply to GeoWeb-created maps, the latter *look* as if they conform to the formal structures of the 'official map' (in a way that counter-cartographers such as the Situationists very clearly rejected). Drawing again on the critical approaches of human geography, we can begin to clarify what sort of structures take their place.

Human bias meets geometric objectivity: cultural appropriations of the GeoWeb

The GeoWeb of direct, georeferenced observation aggregated from the individual to the many has driven new ways of generating cultural content and forming cultural viewpoints. The inherent encoding of human bias, where the common factor is the geometric accuracy of the referenced data point, with no investment on the part of the contributors in any kind of ideological or cultural commons, creates a potent combination for cultural appropriation. Previous literature dealing with academic geographic applications have focused on this, most notably in academic geography, the field where such dichotomies between the 'open' and the 'professional' geographer are most obviously relevant (and problematic). A distinction has been drawn between pedagogically grounded, methodologically based applications of GIS (indeed any such related tool or method), which are the province of professional geographers, and 'fun' applications of neogeography. This distinction been articulated most recently by Leszczynski (2014) as illustrating the difference between the *production of knowledge* and the *acritical gathering of facts*. Haklay (2013) extends this by arguing that GeoWeb-generated maps offer a potentially delusional veneer of faux-democratization. Similarly, Leszczynski focuses on the *geographic* aspects of neogeography, as opposed to the 'newness' implied by the prefix 'neo-'. She argues that the puported 'newness' of neogeographic tools, such as Google Maps, acts as a 'discursive tactic' used by powerful entities (such as Google) to facilitate their appropriation of such technologies for their own (commercial, social or political) ends. Taking a historical perspective, this highlights the distinction made above, between GeoWeb platforms which adhere to – and thus reinforce – the quantitative geographic paradigms of Cartesian place, and the counter-cartographic movements of the twentieth and twenty-first century that set themselves up to disrupt them. In a similar vein, Haklay points out that simply because neogeography platforms increase their numbers of users, this does not equate to those users being empowered, nor does it in some way 'democratize' data or institutions. Both arguments present the GeoWeb (although this is not a term that either amplify) as an acritical mass of connectivity, which acts not in the service of scholarship, but whose very 'openness' renders it vulnerable to exploitation by corporate interests, while puporting to serve a higher public interest.

An even more insidious form of appropriation imbues neogeogeographic data as a result of the demographic nature of the user bases which are predisposed to contribute to them. In her survey of the users behind much of the data behind OpenStreetMap and Google MyMaps, Stephens found that features of likely interest to the demographically dominant groups of OSM users – white men – were likely to be significantly foregrounded, and those likely to be of interest to other groups reduced in visbility. This is not because there are any more of the former or the latter, but due to the nature of the platforms' demographics, and the 'open' processess used to select features for representation, women and ethnic groups found themselves badly under-served, just as they are under-represented in the user bases through (entirely passive) decisions not to participate or contribute (Stephens, 2013).

Disparities in the density and coverage of open geodata can also reflect political and social boundaries. For example, a 2014 study conducted by the author, with the assistance of students from Stanford University's Center for Electronic Spatial and Textual Analysis, into street names in the divided city of Nicosia, Cyprus, represented on OSM, found that 82 per cent of all streets in the Greek Cypriot sector had been assigned names in OSM, as compared to only 32 per cent in the Turkish Cypriot sector. This suggests that there is considerably less engagement with OSM in the Turkish north. One might speculate as to the reasons for this – less broadband coverage in the north, as well as social, cultural and political reasons the population might have on both sides of the border – but the fact remains that thc dataset is, in some way, culturally conditioned by the factors which enable and restrict 'active contribution'.

These debates, conducted mainly in the literatures of human geography, position neogeography as a mechanism for shaping and influencing contemporary society (in the interests or in the image of which organizations or communities is a different matter); and expanding or constraining the choices of individuals, rather than as an analytical tool for better understanding the human record. The same goes for neogeography. Its potential as a methodology, or as a tool, for the analysis of humanities data is less obvious, whether that relates to data of the present day, or of the past. Whatever the arguments around the reductive nature of GIS, for example, it was never doubted that GIS had a role to play as a tool for academic research.

The GeoWeb and the spatial humanities

This discussion shows that neogeography 'in the wild' most likely has little in common with the epistemologies and the approaches of the spatial humanities. The rich diversity of spatial communication on the GeoWeb, detached from the formal structures of 'official' cartography, is inherently prone to political, cultural and demographic appropriation. Spatial humanists who wish to exploit its methods (without exploiting its contributors) must therefore consider how they might 'appropriate' in the place of the multinationals. As critically framed by the human geography literature above, there are two possible areas where this can happen. The first of these is the creation of data and knowledge through geographically oriented academic crowdsourcing, with crowdsourcing interacting with the affordances and practicalities of trans-scalar place on the GeoWeb. The second is 'increased narrativity', through open-ended data exploration as argued by Earley-Spadoni (2017); or an embracing of both analysis and narrative, as argued by Lock & Pouncett (2017). Narrativity, storytelling, the drawing of lines of meaning through data – these are not things that GIS was designed for. This, ultimately, is the challenge of using neogeography in any academic discipline concerned with place, whether that is in the humanities or not: what that critical framework is, and how it can 'anchor' the resulting data and analyses in a way that is academically robust. This is explored further in Chapters 6 and 8.

Both of these possible areas of application involve replacing the acritical structures of neogeography with grounded frameworks for engaging with geodata created outside the academy. This remainder of this chapter seeks to provide a rationalization of the critical frameworks that neogeographic techniques require to form part of the academic spatial humanities, on the basis of these two approaches. It is thus possible to avoid the kinds of critical pitfalls highlighted by academic geographers such as Leszczynski and Haklay. The remainder of this chapter discusses the intersection of neogeography with academic crowdsourcing. The following chapter deals with the emergence of narrativity on the GeoWeb.

Spatial humanities as crowdsourcing

The creation of geographic data by the public for research purposes forms part of the wider phenomenon of academic crowdsourcing, which has been significantly theorized and researched as a methodology in the last few years (Ridge, 2014). Academic crowdsourcing in the humanities is a set of responses by the academy to a world which is increasingly connected and online. As with many other components of the spatial humanities, its historical context is critical: in the mid-2000s, when the concept of crowdsourcing first emerged (Jeff Howe first coined the term in an article in *Wired* magazine in 2006), it referred in the main to the creation and mechanistic processing of large quantities of scientific and commercial data. The best known academic example of this at the time was Galaxy Zoo, which drew in tens of thousands of volunteers to classify images of galaxies from the Hubble space telescope, and attracted a very large (and loyal) contributor base (Bamford et al., 2008). It is significant, and certainly not coincidental, that the GeoWeb, OSM itself (2004), and Google Earth (2005) also began to emerge at this time, as described above.

More recent applications of crowdsourcing however, and especially those in the humanities, seek to engage participants in the process of research itself, rather than simply creating or processing data in a way which requires little or no critical engagement. In such cases, the public can participate not only in the creation and processing of digital research content (for example through the transcription of handwritten documents so that they can be electronically stored and retrieved – see Causer, Tonra, & Wallace (2012)), but also in the design of research methodologies used to investigate that content, and even the research questions themselves. Hedges and Dunn (2017) argue that academic crowdsourcing developed in three phases: *functional crowdsourcing*, corresponding roughly with the period between 2006 and 2010, where tasks set by central research teams are undertaken by anonymous distributed networks providing free labour; *crowdsourcing 2.0*, corresponding again roughly to the period 2010 to 2014, where individual members of those networks begin to display greater levels of autonomy and determination of the creation of content due to the interactive nature of the WWW, and finally *co-production*, where the crowd begins to contribute to the academic research agenda, for example by acquiring and sharing domain knowledge, and disseminating it through activities such as the creation of Wikipedia pages. Central to this is the emergence of so-called 'super contributors', participants in projects who contribute a disproportionate amount of time and/or effort to the project, and the successful cultivation of relationships between members of the professional research community and those individuals. It might be argued that the existence of a successful relationship between the academy and individual contributors is what defines a successful academic crowdsourcing project.

The general phenomenon of 'super-contribution' in academic crowdsourcing, and the interest in academic subject areas which drives it, is what distinguishes 'academic geographic crowdsourcing' from neogeography. Most users of neogeography platforms are not, in and of themselves, interested in the critical affordances of place, space and location in discourse or the environment, in the sense that those engaged with the GIS-driven spatial humanities are.

As noted above, their motivations tend to be intrinsic, they are interested in planning a route, in finding and locating a particular service or facility, or in contributing to georeferenced resources in a very specific and localized manner. In other words, they are not 'active contributors' in the terminology of See et al. (2016), or 'super-contributors' in the terminology of Hedges & Dunn (2017).

According to the study of See et al. (2016), who surveyed a range of terminologies used to describe the collective and collaborative acquisition of geodata, projects which foster the development of 'super contributors' are 'actively contributed'. This, for example, would include citizen science-type projects, where members of the public record the locations of sightings of a particular species. Most importantly for the present discussion, it means that the participants/contributors have made a proactive decision to participate in the project, and contribute some form of data to it; and that there is a guiding rationale for how that data was created. Returning to the question of motivation, this also becomes much clearer. As Hedges & Dunn (2017) found, subject area motivation is a key factor for such 'super contributors'. They are motivated by interest in a particular subject, and to engage with content related to that subject ('asset' in the terminology of Hedges & Dunn). A key question therefore is what manner of geodata is produced during this process of 'super contribution'. In contrast, according to the analyses of neogeography cited above, the *absence* of a critical focus is the key feature of neogeography mapping platforms, even though these are often couched in terms of allowing users to have control of transactional data they create in their daily lives. As with the distinction between the neogeography as opposed to counter-cartography, there is no guiding principle or rationale for collectivized geodata, in the way that geographic crowdsourcing provides. There is thus a key distinction between geographically oriented academic crowdsourcing, and the 'wild west' of neogeography.

What might such a guiding principle, the defining feature of 'academic geographic crowdsourcing' be? Certain kinds of academic crowdsourcing, those which perforce require a substantial connection between individual participants and professional researchers, have particular places as their focus – and this focus binds the project. A good example of this is the Cambridge Community Heritage project, in which archaeologists worked with residents of four rural communities in Cambridgeshire on small test-pit excavations in their local areas (Lewis, 2015). The archaeologists coordinated the siting of pits, but the residents did the digging, making discoveries of artefacts, which the project team then identified and dated. While this produced a substantial quantity of archaeological data in its own right, more importantly it facilitated contact between perspectives of the area's history in and the present-day community. It thus enhanced and deepened the latter's sense of history, and sense of place. The project's leader notes in her conclusion:

> Viewed in the context of deep mapping, it is salient that community test pit excavation projects create performative and perceptual links between

> present and past communities, as people are connected with both the process and product of following horizontal and vertical pathways of discovery about their own locales, while simultaneously generating new evidence which can be mapped within its spatio-temporal context and aggregated to advance wider understanding of changing places in time. (Lewis, 2015: 413–14).

This is perhaps an extreme example, but it illustrates geographic crowdsourcing, with a particular space and location at its heart, that is critically curated and directed in the service of increasing understanding of and knowledge of place, and of increasing people's lived connection with it. This is the opposite end of the spectrum to the open, acritical standards-based geographies of OSM and Google MyMaps. It is furthermore not vulnerable to the threats of cultural, commercial and political appropriation highlighted by Lezczynski and Haklay.

Academically crowdsourced geodata and practices can be further distinguished from what Goodchild and others would call 'VGI' or 'neogeography' by the motivations of those contributing the data, and the kinds of relationships they have with professional and/or institutional structures – in this case university researchers who are responsible to institutional frameworks, ethics committees, professional standards, etc. However, it is less useful to see such projects activities as belonging to hard and fast categories; rather they exist as part of a methodological spectrum.

Employing the idea of the spectrum allows us to distinguish between critically detached analysis of collectively produced geographic data, which Fink (for example) refers to as being 'a significant part of cartography and thus an important subject of cartography research' (Fink, 2011: 1), and bodies of collectively produced geodata (such as that of the Cambridgeshire project) which elucidates answers to questions about the human past and the human present. This acknowledges that the geodata produced through academic crowdsourcing is a *means* to knowledge production, whereas the content produced by neogeography should be seen as a *subject* of critical cartographic or geographic investigation.

Base maps

This does not mean that lower-level resources and datasets resulting from what, in the framework above – and following the analysis of See et al. (2016) – one might refer to as generic neogeography or VGI – cannot be *used* in applications which support spatial humanities research. One obvious example of this is the base maps on which many spatial humanities projects rest. Of 76 such projects surveyed by Luchetta (2017), for example, more than 50 relied on base maps such Google Earth, ArcGIS Online, and OpenStreetMap (ibid. 8). The last of these, OSM, is a world base map created through crowdsourcing contributions from an open user community

(Haklay & Weber, 2008). In the light of this separation between content and analysis, it is clear that OSM should not be considered an 'academic crowdsourcing project': a large majority of its uses are, simply, not for academic purposes.

OSM is currently one of the largest and most visible VGI base maps, and it has thus attracted a significant amount of critical attention. This attention has focused on the nature of its crowdsourced contributions (Budhathoki & Haythornthwaite, 2013), its quantitative accuracy as compared with 'official' base maps such as the Ordnance Survey, and the way in which neogeography acts, or does not act, as an agent of 'democrataization'. Most of these studies concern the user/contributor base itself. Budhathoki and Haythornthwaite develop a distinction between 'heavyweight' and 'lightweight' contribution to the platform, where the latter rely on individual interventions, equivalent to using a 'like' button on social media, requiring little or no prior technical knowledge, and no geographic knowledge beyond that needed to capture a single observation in the immediate environment and transfer it to OSM by means of one of the user-facing in-browser editing tools, such as Potlatch. This mode of contribution, corresponding to 'mechanical' tasks in the typology of Hedges & Dunn (2017), is mainly characterized by the fact that they are carried out by non-communicating individuals engaging with single individual pieces of data at a time. A 'heavyweight' contribution on the other hand requires more investment of time, knowledge and expertise and, especially, it is made in the context of the OSM community – where it is rewarded by recognition of the weight of the contribution in the community's social structures, and the impact it has on the aims of the project.

Interestingly in this context, the case of OSM in recent debates has highlighted the limitations of presenting crowdsourced geographic data as data only, rather than as a resource that can be used for a range of facilitated purposes by the user community. Most notably, Wroclawski (2018) regarding the title of a blog post, explains '[w]hy is OpenStreetMap in serious trouble'. He states, as one of the main barriers for OSM uptake, that the emphasis is too much on the provision of data, rather on the provision of services that make that data useful and usable:

> The first problem that I feel plagues OSM is that the OpenStreetMap Foundation views the mission of the project to provide the world a geographic database, but not geographic services. OSM gives people the tools to create their own map rather than offering them a simple, out of the box solution. … And even if [smaller users] do use our data, their engagement is through a third party, rather than directly with us (Wroclawski, 2018)

This adds to the critical concerns raised by Stephens and others about the way in which open geodata is created. Not only does the absence of frameworks to regulate the non-technical (and thus cultural, social and political) creation of geodata lead to the kinds of gender and racial biases noted above, the provision of the dataset as a raw mass of geo-information places it out of the reach of organizations which might press its usage and development in different directions. As noted above, the motivations of such organizations are easier to identify and label than those of individuals, and are thus more accountable for the social and political forms which the data they create take.

Conclusion

The intersections between base maps and applications, content and analysis, amateurism and professionalism, automated aggregation and expert intervention described above highlights further the dichotomy between technical infrastructure and cultural practice with which human geographers have approached neogeography. It also mirrors views of geography as being hybridized and holistic, transcending distinctions such as human/physical, quantitative/qualitative, activism and academia, and so on (Sui & DeLyser, 2012: 112–13). The move towards 'synthesis and holism' which Sui and DeLyser describe has impacted the spatial humanities in parallel ways.

As with human geography, this involves rejecting binary classifications of epistemology, and embracing the concept of the spectrum. And this hints at an answer to the question posed at the start of this chapter, of what place the GeoWeb occupies in the spatial humanities: it is, with a critical framework based on the intersections listed above, a vehicle for the production and communication of spatial narrative. Caquard characterizes this content–analysis dichotomy as a distinction between 'grid maps' and 'story maps' (Caquard, 2011: 6), qualifying the distinction by arguing that it becomes unmeaningful, since '[t]his type of application stimulates the production of spatial narratives by making them easy to map and distribute, and simultaneously restricts them through the framework provided by the base map' (ibid.). The key point here is how the production of spatial narrative is 'restricted' by the base map, as is the creation of neogeographic geodata unrestricted by anything but technology and data standards. The implication of Caquard's analysis is that this restriction comes primarily in the form of formalization and quantification, the reduction of discursive data to points, lines and polygons, latitudes and longitudes. However, the observations above about the political appropriation of neogeography, and the biases that exist under its representational shell, mean that any spatial narrative that is created using neogeographic data is shaped by its inherent cultural biases as well as its formalistic, un-narrational structures.

The spatial humanities thus require that some form of order be brought to the GeoWeb. This order can replace the (commercially oriented) direction provided by national mapping agencies, and the corporate exigencies of entities such as Google – all now the 'authoritarian' and 'top-down' providers of Goodchild 2007. But the popularity of OSM among app and web service developers and academic researchers who need such base maps (Mooney,

2015) brings problems. The allure of 'openness' in terms of what can be contributed and by whom combines with the allure of being free to use; but without problematization, open geodata risks bringing the latent biases inherent in it into the academic discourse.

Open cartography's appeal for end users is further enhanced by the fact that its levels of geometric accuracy stand up relatively well to comparisons with mapping agency maps. Comparative studies of OSM and national mapping agencies, such as the UK's Ordnance Survey, have shown that in terms of quantitative accuracy, it compares extremely well (Haklay, 2010), notwithstanding issues such as the 'digital divide', which result in coverage of areas with lower levels of broadband coverage getting less dense feature coverage. This is particularly the case with major features such as motorways and A-roads, which attract significant levels of daily usage and attention in the landscape.

Part of the reason for this is the fact that the kind of active contribution on which academically crowdsourced geodata clearly depends is conditioned by *scale*. OSM self-consciously aims to be a *world* map covering the totality of the planet's surface (Haklay & Weber, 2008). Active contribution, then, to the contributory tasks which make its accuracy comparable – with the caveats above – to agencies such as the Ordnance Survey, thus involves undertaking a generic task, which is identical in terms of the protocols, interfaces and data produced (including georeferencing of the data to WGS84 grid standards), wherever on the world one is undertaking it. However, one class of project which falls in to the rubric of the spatial humanities deals with areas which are relatively small and more specific in their geographic scope, and thus rely specifically on contributions from individuals with context-specific local knowledge. Such projects depend on active contributions to tasks which are similarly designed and executed for that area, and are thus not to generic, and digitally replicated and mediated, but rather to discursive and reflexive bodies of knowledge. This, in turn, depends heavily on locally based users having – and wanting to contribute – reserves of highly specific local knowledge about a particular subject in a particular location. They are thus actively contributing to a larger exercise, but – crucially – it is not an activity which could be (or needs to be) sustained over any larger geographic scale.

References

Bamford, S. P., Nichol, R. C., Baldry, I. K., Land, K., Lintott, C. J., Schawinski, K., … Vandenberg, J . (2008). Galaxy Zoo: the dependence of morphology and colour on environment. Available online: https://doi.org/10.1111/j.1365-2966.2008.14252.x

Bednarz, S. W. & Kemp, K. (2011). Understanding and nurturing spatial literacy. *Procedia – Social and Behavioral Sciences,* 21, 18–23. Available online: https://doi.org/10.1016/j.sbspro.2011.07.004

Budhathoki, N. R. & Haythornthwaite, C. (2013). Motivation for open collaboration. *American Behavioral Scientist*, 57(5), 548–575. Available online: https://doi.org/10.1177/0002764212469364

Caquard, S. (2011). Cartography I: Mapping narrative cartography. *Progress in Human Geography*, 37(1), 135–144. Available online: https://doi.org/10.1177/0309132511423796

Causer, T., Tonra, J., & Wallace, V. (2012). Transcription maximized; expense minimized? crowdsourcing and editing The Collected Works of Jeremy Bentham. *Literary and Linguistic Computing,* 27(2), 119–137.

Earley-Spadoni, T. (2017). Spatial History, deep mapping and digital storytelling: archaeology's future imagined through an engagement with the Digital Humanities. *Journal of Archaeological Science*, 84,95–102. Available online: https://doi.org/10.1016/j.jas.2017.05.003

Fink, C. (2011). Mapping together : On collaborative implicit cartographies, their discourses and space construction. *Journal for Theoretical Cartography*, 4,1–14.

Ford, H. & Graham, M. (2016). Semantic cities: Coded geopolitics and rise of the semantic web. In R. Kitchin & S.-Y. Perng (eds.), *Code and the city*, pp. 200–214, Routledge. (JANUARY), 1–37. Available online: http://papers.ssrn.com/abstract=2682459

Goodchild, M. F. (2007). Editorial : Citizens as voluntary sensors : Spatial data infrastructure in the World of Web 2.0. *International Journal*, 2,24–32.

Haklay, M. (2010). How good is volunteered geographical information A comparative study of OpenStreetMap and Ordnance Survey datasets.pdf. *Environment and Planning B: Planning and Design*, 37,682–703.

Haklay, M. M. (2013). Neogeography and the delusion of democratisation. *Environment and Planning A*, 45(1), 55–69. Available online: https://doi.org/10.1068/a45184

Haklay, M., Singleton, A., & Parker, C. (2008). Web mapping 2.0: The Neogeography of the GeoWeb. *Geography Compass*, 2(6), 2011–2039. Available online: https://doi.org/10.1111/j.1749-8198.2008.00167.x

Haklay, M. & Weber, P. (2008). OpenStreetMap: user-generated street maps. *Pervasive Computing, IEEE*, 12–18. Available online: http://discovery.ucl.ac.uk/13849/1/13849.pdf

Hedges, M. & Dunn, S. (2017). *Academic crowdsourcing in the humanities: crowds, communities and co-production*. Elsevier.

Hill, L. L. (2009). *Georeferencing: The geographic associations of information.* MIT Press.

Jiang, B., & Yao, X. (2006). Location-based services and GIS in perspective. *Computers, Environment and Urban Systems*, 30(6), 712–725. Available online: https://doi.org/10.1016/j.compenvurbsys.2006.02.003

Junglas, I. A., & Watson, R. T. (2013). Location-based services. *Communications of the ACM*, 51(September),65–69. Available online: https://doi.org/10.1145/1325555.1325568

Lake, R. & Farley, J. (2009) Infrastructure for the Geospatial Web. In A. Scharl & K. Tochtermann (eds.), *The geospatial web. Advanced information and knowledge processing*. London: Springer, pp. 15–26.

Leszczynski, A. (2014). On the neo in neogeography. *Annals of the Association of American Geographers*, 104(1), 60–79. Available online: https://doi.org/10.1080/00045608.2013.846159

Lewis, C. (2015). Archaeological excavation and deep mapping in historic rural communities. *Humanities*, 4(3), 393–417. Available online: https://doi.org/10.3390/h4030393

Lock, G.. & Pouncett, J. (2017). Spatial thinking in archaeology: Is GIS the answer? *Journal of Archaeological Science*, 1–7. Available online: https://doi.org/10.1016/j.jas.2017.06.002

Luchetta, S. (2017). Exploring the literary map: An analytical review of online literary mapping projects. *Geography Compass*, 11(1), e12303. Available online: https://doi.org/10.1111/gec3.12303

Mooney, P. (2015). *An outlook for OpenStreetMap* (pp. 319–324). Springer. Available online: https://doi.org/10.1007/978-3-319-14280-7_16

Offen, K. (2012). Historical geography II: Digital imaginations. *Progress in Human Geography*, 37(4), 564–577. Available online: https://doi.org/10.1177/0309132512462807

Perkins, C. (2007). Community mapping. *The Cartographic Journal*, 44(2), 127–137. Available online: https://doi.org/10.1179/000870407X213440

Ridge, M. M. (2014). *Crowdsourcing our cultural heritage.* Ashgate Publishing.

Scharl, A. (2007). Towards the geospatial Web: Media platforms for managing geotagged knowledge repositories. In A. Scharl & K. Tochtermann (eds.), *The geospatial Web – How geo-browsers, social software and the Web 2.0 are shaping the network society.* (pp. 3–14). Available online: https://www.geospatialweb.com/sites/geo/files/sample-chapter.pdf

See, L., Mooney, P., Foody, G., Bastin, L., Comber, A., Estima, J., … Rutzinger, M. (2016). Crowdsourcing, citizen science or volunteered geographic information? The current state of crowdsourced Geographic Information. *ISPRS International Journal of Geo-Information*, 5(5), 55. Available online: https://doi.org/10.3390/ijgi5050055

Stephens, M. (2013). Gender and the GeoWeb: divisions in the production of user-generated geographic information, *GeoJournal* 78(6), 981–996.

Sui, D. & DeLyser, D. (2012). Crossing the qualitative-quantitative chasm I: Hybrid geographies, the spatial turn, and volunteered geographic information (VGI). *Progress in Human Geography*, 36(1), 111–124. Available online: https://doi.org/10.1177/0309132510392164

Sui, D., Goodchild, M., & Elwood, S. (2013). Volunteered geographic information, the exaflood, and the growing digital divide. In *Crowdsourcing geographic knowledge* (pp. 1–12). Springer.

Turner, A. (2006). *Introduction to neogeography.* O'Reilly Media, Inc.

Wroclawski, S. (2018). Why OpenStreetMap is in serious trouble — Emacsen's Blog. Retrieved 17 July 2018, Available online: https://blog.emacsen.net/blog/2018/02/16/osm-is-in-trouble/

Zook, M. A. & Graham, M. (2007a). Mapping DigiPlace: Geocoded Internet data and the representation of place. *Environment and Planning B: Planning and Design*, 34(3), 466–482. Available online: https://doi.org/10.1068/b3311

Zook, M. A. & Graham, M. (2007b). The creative reconstruction of the Internet: Google and the privatization of cyberspace and DigiPlace. *Geoforum*, 38(6), 1322–1343. Available online: https://doi.org/10.1016/J.GEOFORUM.2007.05.004

6 Spatial narrative

Introduction: the idea of spatial narrative

We navigate, understand and interpret the world via our perceptions and mental maps of it, and much of these perceptions and mental maps are based on secondary spatial and place information. A theme running through the previous chapters concerns the different structures through which this information reaches us: how it is communicated and socialized, and the contexts of those structures. Socialized communication of place is always bound up with the affordances of communication media: we have traced an historical arc from the early modern period and the affordances of movable type, to more recent periods of history when the Internet assumed primacy as the means for communicating information, including information about, and representations of, place. The notion of 'spatial narrative' has also emerged at various points. This chapter examines the idea of spatial narrative as a way of viewing place-making in the humanities in the digital age.

Narrative is what makes the communication of any form of descriptive statement, spatial or otherwise, meaningful. As noted in Chapter 1, most humanities disciplines are framed by the epistemological nature of their content, and so a key part of those working with such content is to tell meaningful stories about it. Throughout the history of spatial meaning-making, the distinction between spatial *data* and spatial *information* has remained irresistibly irreconciled, though the dichotomy of what can be experienced and what can be measured. Spatial *data* is information about space and place rendered into quantitative form via a process of georeferencing or geocoding, such as geodetic latitude/longitude points which allow 'place' as thus abstracted to be replicated exactly and frictionlessly on the recipient'(s') device(s); as binary groups of information communicable at the speed of light via electrons. Spatial *information* is described qualitatively and communicated through analogue channels: it is subjective, prone to interpretation and alteration as it passes between different individuals, and through levels of reception. The geographic and the chorogaphic, the datalogical and the infological, the quantitative and the qualitative continue to vex the humanist whichever technological and communicative systems he or she is moving through, yet both can form part of a *narrative* which makes descriptive statements meaningful.

We may therefore approach this from a theoretical perspective, in which the transmission of information from one human context to another is defined as *spatial narrative* (Bodenhamer, Corrigan, & Harris, 2015). Narrative, a term which has arisen frequently in the preceding chapters, is usually constructed in terms of connected events and/or characters, which come together and take on a logical significance (or meaning) through being linked by a plot or plots. Yet if we qualify narrative as being 'spatial', it becomes a potential bridge between the qualitative and quantitative approaches, where information and data are linked by place, not by plots. A spatial narrative is a meaningful statement or event, or set of linked statements or events (narrative components, equivalent to events), which take on a homogenous significance as a result of shared understanding of a place or places, and the different meaning(s) that can be attached by different entities to humanistic place.

In the process of sharing and encoding, extrinsic empirical truths relating to the physical landscape, and the intrinsic truths of human perception of it become linkable, all the more so since these structures have become ever more sophisticated with the refinement of mapping techniques and communication structures in the modern era, including (for example) the introduction of photogrammetry into topographic mapping from the 1950s (Collier, 2002) and, subsequently, satellite mapping. Through a process of critical quantification (see next chapter), for example the construction of WWW-readable gazetteers, it becomes possible to link both information and data. This begets an increasingly complex relationship between the narrative elements of a map which lay claim to spatial accuracy, and those which represent place in a 'human' or 'cultural' sense. This distinction is described in the literature in various ways. Caquard, for example, talks of the distinction between a 'grid map' and a 'story' map (Caquard, 2011), whereas Leszczynski, as we have seen, uses the term 'datalogical' to describe mapping which conveys correct geometric detail, and 'infological' for that which describes space in unmeasurable human terms (Leszczynski, 2009). The rendering of points of spatial communication and, subsequently, interpretation into narrative events via processes of critical quantification means we can we can measure (as opposed to map) a region, say an English county, in an 'accurate' Cartesian manner, such as one employed by the Ordnance Survey, and at the same time express emotional and 'psychogeographcial' interpretations of the same county as perceived by human sensory experience.

Spatial narrative framed as means towards understanding, rather than a definition of content type, thus has the advantage of not requiring that space to be quantitative or qualitative (or infological or datalogical, after Leszczynski), but both at the same time, forming networks of different narrative components that can be either. Such a component can appear in any of the forms of humanistic discourse discussed in previous chapters – maps, text, images, performative spaces and so on. Spatial narrative networks are *transmedial.*

Here it is argued that spatial narratives expressed in this way fall into three broad categories: they can *facilitate change*, they can *document personal experience*, and they can *assert power*.

Framing transmedial spatial narratives as networks which cross different humanities disciplines and content types reflects thinking in other areas where narratives which are not necessarily linear have proved valuable for meaning-making. The material record, for example, and its understanding and reception through narrative, or narrative-like knowledge structures, has been addressed through examination of typology-based chronological sequences (Hodder, 1993), through objects (Gosden & Marshall, 1999), through human experience (Brück, 2005) and through archaeological publication as a literary topos (Pluciennik, 1999). The role of museums and cultural collections in the formation of narratives has also been extensively considered (Rowe, Wertsch, & Kosyaeva, 2002; Moser, 2010). These are spaces for narrative creation, in which one narrative can result from many 'voices'. It touches on multi-authorship of narratives by professional archaeologists and curators, how they are formed and reformed, and are challenged, by exposure to the public. Another advantage of the idea of spatial narrative which crosses discourses is that – like these other kinds of narratives – it can have either one narrator or many.

The idea that a spatial narrative represents a network possibly containing both experiential and measurable statements is also useful. Events within each of the types of spatial narrative explored below have more in common with 'events' mapped into a conceptual structure like the CIDOC CRM, rather than events defined by dramatic significance normally associated with the term. Such 'events' concern the relationships between places, people, objects and other kinds of entity across different media: these can be text, in human embodiment, on a map, on a schematized diagram, or through the spoken word. A spatial narrative is a system of information which causes, forms or records any human idea of place and its significance. The relations which constitute it can be formed by oral tradition, through gesture, or they can exist as memory. Human-formed landscapes allow for the construction of spatial narratives in different ways, whether these be narratives of power or separation, narratives of association, memory, or narratives of embodiment, involving the restriction or enabling of movement. As noted, the component parts of a spatial narrative can be both quantitatively measurable and qualitatively describable. So if, as argued in the previous chapter, WWW-mediated communication of place collapses the distinction between *chorography* and *geography* in an acritical and un-problematized manner, spatial narrative can provide a critical framework for re-defining them. Furthermore, a holistic approach to spatial narrative can help overcome the dichotomy between the quantitative and the qualitative place which has framed the uptake of GIS and digital mapping technologies in the humanities, leading – hitherto – to an epistemic distinction between spatial narrative (of the kind enabled by Web 2.0, and neogeographic approaches) and analysis, as underpinned by quantitative GIS investigation.

Spatial narrative is thus a systemic means by which spatial knowledge is first shared, and then understood, by two or more entities. The idea of sharing geographic data from one entity to another in a form which is understandable to multiple readers has been of fundamental significance to mapping for centuries. The principal significance of Ptolemy's *Geography* was its introduction and use of an independent and tabular latitude/longitude system, a structure that was sharable and replicable. The disconnection between place and human reading of place also exists in time: a (paper) map is both a freeze-frame representation of a section of the earth's surface at a particular moment, and encoding of the cartographer's decisions about what to include and define, and how to bound and abstract the physical complexity of the world. Always, the spatial narrative is framed by the communication network or networks in which it resides. The processes of mechanically transferring data from device to device contrast with that of one person expressing information verbally to another, and indeed with ekphrastic networks of reading, as described in Chapter 4.

These distinctions matter in the creation of spatial knowledge in the humanities. Archaeology, as we have seen, is a discipline which relies extensively on the publication and communication of quantitative data such as the find spots of artefacts, the measurement of features, and the precisely enumerated expression of evidence, such as the number of objects in a particular assemblage. As outlined in Chapter 3, the assumption that historical space can only be narrated through quantification and enumeration framed much of the thinking of archaeology as a discipline in this modern form, and led to the idea that human behaviour consists of processes which are replicable, governed by universal laws, and are thus predictable.

These are examples of how spatial data or spatial information becomes meaningful through communication, and what form that meaning takes. The last chapter reflected on the communication structures that researchers in the academy work with to create new forms of spatial knowledge in conjunction with non-academic groups, in the course of geographic crowdsourcing projects. The meaning produced by these projects takes the form of grounded academic knowledge (in the case of the Cambridge Community Archaeology project, knowledge about the localized past of the four villages in the survey), expressed in the form of published peer-reviewed papers, and as less formal outputs such as increased knowledge of (again, for example) archaeological or historical features among local communities. But this is only one kind of meaning – one kind of spatial narrative.

As scholars of critical cartography have long noted, maps are not value-free vessels for the visualization and transmission of geodata, rather they are intensely political artefacts of theory, perspective and ideology (Crampton & Krygier, 2006). This is also a fact noted by scholars in disciplines heavily engaged with the spatial humanities such as archaeology (Lock, 2010), and those who have highlighted the 'implicitness' of cartographic messages (Fink, 2011). Spatial narratives can thus be categorized according to the type of meaning they convey. They are agents of explanation and argument, which weave together both quantitative and qualitative information. They are points of view about a human subject or discipline which are backed by evidence and add to knowledge, which makes a case in order to persuade. In the cases of geographic crowdsourcing, these are academic arguments, which arise from a collaborative mediation of professional and amateur perspectives and effort, through (usually) digital methods enabled by the Internet. The spatial narratives of the early modern chorographers referenced in Chapter 2, while being propagated by the pre-Internet means of movable type and printing, nonetheless dealt with position-based arguments, around the spatiality of emergent nationalism. With writers such as Camden, we saw a clear exposition of the Britannia as a national (and nationalistic) entity within Christendom (itself a supra-national spatial entity) moulded in the image of God. This is a narrative, communicated through print networks to a dominant section of contemporary society (the educated, literate, elite male section); and it does not have to substantiated or evidence based.

Spatial narratives can also be agents of power, or at least tools for the powerful and/or the enfranchised. One of the consequences of the opening up of spatial communication infrastructures online is the swift aggregation of individual subjective points of view, to the point where they become narratives which regulate and contain users' perception and appropriation of place – as demonstrated by the examples of neogeographic platforms being unconsciously assumed by their dominant user demographics. What these disparate examples share is that they are all spatialized perspectives formed through communication and common understanding which can be both shared by different entities, albeit from different perspectives (e.g. male and female in Stephens's examples discussed in the previous chapter).

Spatial narrative and Plato's Cave

Spatial narrative framed in this way, facilitated by technologically enabled networks, forms much of our view of the world. It is, in other words, the primary lens which mediates our knowledge of much of the world that we have not, and will never, experience first-hand with our own senses. Most human beings, for example, will never know first-hand (i.e. chorographically) what it is to physically experience Antarctica. However there is a vast array of spatial narratives, of the types described above, which explain events that happen in that environment, which facilitate change, which document personal experience, and which seek to persuade (think of the corpus of materials which draw on Antarctican platial imagery to make arguments about, for example, climate change). This allows us to place spatial narrative in a broader theoretical framework. In a sense, spatial narrative is therefore a transmedial and multimedial expression of Plato's Cave. Plato's Cave is a metaphorical scenario famously put forward by the philosopher to illustrate the distinction between sensory experience and reality. In the Cave, a group of prisoners are chained in a row, facing the far wall of the cavern. They have been there since birth and have never known or seen the outside world. Behind them is a fire, and in the space between move figures, and the figures cast shadows on the far wall. Thus, the only reality that the prisoners have ever known are the shadow images. Plato uses this image to argue that only by engaging with reality with critically trained faculties (such as those possessed by a trained philosopher) can we understand it; most people go through life, like the chained prisoners, relying for our view of reality on the shadow images of secondary media (Wright, 1906).

Plato's Cave has been used before to illustrate the impact of rapidly evolving technology in mediating and conditioning (and in some ways regulating) social views of the world. In her 1978 analysis of the role of photography in contemporary society, Susan Sontag appropriated the allegory to illustrate the usurpation of reality by photographic perception, at a time when the camera was becoming relatively widespread and affordable (she describes photography as 'almost as widely practiced an amusement as sex and dancing'). Sontag writes: 'Reality has always been interpreted through the reports given by images; and philosophers since Plato have tried to loosen our dependence on images by evoking the standard of an image-free way of apprehending the real' (Sontag, 1978: 153). Much the same can be said of the ways in which we interpret and construct the world through spatial narrative: the spatial media, grammars and practices of information-bearing artefacts such maps, text, the Internet and television are our sole frameworks for seeing and 'experiencing' most places. Communication and travel technology, from the telegraph, to the steam, internal combustion engine and, more lately the WWW and Internet, vastly expanded humans' exposure to spatial information and data, at a rate that simply could not be matched by human experience of the real/physical world. The digital revolution has thus greatly expanded the shadow imagery available to the prisoners in the Cave – spatial narratives about places we have never seen for ourselves – and also it allows us to problematize the receptive barriers between the present and places we can *never* experience, for example that described in an historical manuscript. And just as Sontag's image-cave acts as an agent and as an environment for the creation of human identity, so spatial media conditions the spatial identities of nations, regions, communities and individuals.

The data and information on the walls of Plato's Cave are thus the basis through we come to see, interpret and construct our sense of place and space in most of the world. In order to provide a critical framework to understand this process further, we need to explore further the three categories of spatial narrative introduced above: narratives of change, narratives of power, and narratives of experience. The last of these will segue into the next chapter, which deals with the ways in which digital environments relate to human mobility, which is of course closely related in concept. As with most other attempts to categorize approaches within the spatial humanities, these aspects represent a spectrum rather than a set of discrete categories. Combined with an understanding of what is meant by a spatial narrative, they provide a sound way of navigating the distinction between quantitative and qualitative material, and spatial data and spatial information. In the discussion that follows, I draw on a series of examples of the production and elaboration of spatial narratives from historical humanities sources, and in the conclusion reflect on how digital approaches to the spatial data/information divide can help us to research them.

Narratives of change

The Greek-centric nature of Homer's Shield of Achilles, discussed in Chapter 4, highlights the subjective character of mapping and spatial narrative in presenting explanations of particular views of the world. Heffernan's analysis (Heffernan, 1991) highlights the importance of temporality in this: narrative must, perforce, explain *change over time*. As a set of linked, meaningful statements whose basic structure is geographical, a spatial narrative must relate to the description of events which change a place from one type in to another. As well as being spatially bounded, such events require temporal distinction in order to be both tellable and eventful. With spatial-temporal events framed in such a way, one may, for example, map (spatially and temporally) a series of 'events' which connect Baker Street in London in 2018 with the fictional construct of Sir Arthur Conan Doyle. Such events might include Conan Doyle's act of writing a scene set in the street, a set of the fictional events he describes there, and – if applicable – the reader's or viewer's own experience of the contemporary place. Going back in time, before the fictional event was written and set, one could define the act by which it was named 'Baker Street', and the literal spatial boundaries of that place (if this is possible). All of these would document changes, events, bounded by the same place, which also represent modifications in the way that place is seen by different actors at different times.

From this brief worked example, it is clear that linear structures do not lend themselves well to descriptions of place which link spatial references in literary, archaeological, historical, or performative discourses, with places as we see them ourselves. Spatial narratives are nonlinear. However, digitization of such discourses makes humanistic place subject to the standards and platforms of the contemporary WWW/Internet/GeoWeb, which – being aspatial, as discussed previously – enforce a linear 'click by click' pathway through the data. Therefore, digitization of humanistic material, or the investigation of born digital humanistic material, must be seen as only one way in which changeful spatial narratives of history and archaeology can be created in the present day. Spatial narratives can manifest themselves in the user's office or on his or her desktop, but they cannot be detached from physical location. For example, in their discussion of spatial narrative, Azaryahu and Foote frame major historical sites as being 'dedicated to the cultural production of the past' (Azaryahu & Foote, 2008: 179), a process which, in itself, implies that they change one period's perception to another. That change in perception is enacted at single points within the site, through managed or regulated pathways through (or around) it, and through narratives embodied over large regions (ibid.). As Azaryahu and Foote explain, this process in itself involves a process of temporal separation, but there are numerous other separations implied as well: present from past, curator from visitor, inside from outside, space in which visitors are permitted to be or traverse, and spaces which they are not.

Narratives of power

Whatever the technological or standards frameworks, neither spatial data nor spatial information encode ground truth, but rather subjective human views of it. This includes deliberate bias and distortion of these encoded units of information enacted in order to express or assert power. From at least the seventeenth century, maps have been employed as vessels for political, social and cultural arguments some of which, by logical extension, overpower other views. For example, in 1603/4 the British cartographer John Speed (1552–1629) published his *Map of England, Wales and Ireland*, a detailed cartography of the British Isles, alongside an image of the newly crowned James I. The image is placed above a genealogical tree of English rulers dating back to the Norman Conquest of 1066, and in the context of the crests of numerous notables of the realm. The combination of 'modern' [=contemporary] views of the world with genealogical tables produces a clear spatial narrative: James I's rule of the mapped lands to the left is legitimized by what has gone before (Barber & Harper, 2010). This was a device well-known to Renaissance and post-Renaissance cartographers who wished to ground their maps in their political import. In his map of the Roman Empire, the *Romani Imperii Imago* (1579), Ortelius places a lineage diagram of the kings of Rome in the bottom right-hand section of the folio. For cartographers of this period, empire, territory and legitimacy were interchangeable.

Similar spatial narratives take on a direct visual representation of national symbolism, as in the Leo Belgicus maps of the Netherlands, which include modern Holland and Belgium. In these maps, the outline of the latter countries are visually manipulated into the shape of a lion, an animal symbolizing the strength and legitimacy of the Belgian imperial project. The close relationship between maps and imperial imagery is also a motif of British imperial cartography of the nineteenth century. In 1886, Walter Crane published his *Imperial federation Map Showing the Extent of the British Empire*, quintessential example of imperial cartography which shows Britain's overseas dominions, coloured in the standard pink, with colonial shipping routes, and tables of statistics showing the produce of each region. Around the map is an array of colourful figures of stereotyped colonial characters – the Native American, a maharaja on an elephant, eroticized eastern women, and the colonizers themselves in the form of British soldiers and sailors (Biltcliffe, 2005).

Significantly, Crane himself was of a socialist outlook, and the narrative, discussed by Biltcliffe, is one of the dignity of labour, of the Empire as a benevolent federation (one of three words emblazoned on banners at the top of the map, the other two being 'Freedom' and 'Fraternity') of nations in which the right to be productive is safeguarded by British stewardship.

In these examples of spatial narrative, the subjective (and partisan) message is clear enough although, in the case of the Crane map, a rather deeper reading is required, as discussed by Biltcliffe. However, the employment of narrative techniques is often more subtle, and requires a deeper and more discursive reading. For example, place-name labelling on maps encodes significance with etymological, semantic or stylistic means. Lock's (2010) argument that map labelling explicitly attempts to avoid subjective interpretation in the symbolic representation it offers, must be questioned. Indeed, a cursory consideration of the history of map labelling suggests that, especially in historic maps, the style and typefaces of labelling are frequently used to encode, explicitly or implicitly, spatial classifications and therefore narratives of hierarchy. This may simply be a useful device for differentiating between important and less important settlements. Early Ordnance Survey maps of the UK, for example, used different fonts and typefaces to distinguish between county, parliamentary and municipal boroughs, and between smaller kinds of administrative district. This has the effect of categorizing them visually, and thus of (re)enforcing a shared understanding of the relative significance of different places in early twentieth-century England, where significance might come in the form of population, being the seat of a political or administrative entity, or even simply from being a certain kind of place, such as an ancient site – the convention of using Gothic script to label sites of antiquity persists in contemporary Ordnance Survey mapping.

One can use labelling to trace replication throughout historical sources, using almost any place-name as an example. One might take an example from the United Kingdom: Holbeach, a small town in Lincolnshire, which can be found on the 1903 and 1923 Ordnance Survey series maps, with the lettering captialized, highlighting the town's significance as compared to neighbouring conurbations. Holbeach was the ancestral home of William Stukeley (1687–1765), whose personal notebook contains a sketch plan, published by the Ordnance Survey in 1925, of the area of his father's house in the town. The label 'Plan of my father's dwelling at Holbeach' personalizes it, and hints a personalized instance of the geography. This background information might be regarded as the equivalent to a critical apparatus of an edition of a text: a 'story map' approach, in the terminology of Caquard (2011); but at the same time it is a base map enriched and 'deepened' by text in practice. Maps are therefore 'deep' information objects. However, although a map is a deep and socially contextualized representation of human perception of humanistic place (Lock, 2010), it is one which is framed by the sharable structures of cartography, namely grids, references, latitudes, longitudes, the conventions of scale and the linguistic structures of toponymy. These do not represent personal or relational structures of place (in the way that medieval maps did), but place maps clearly in the category of devices used for functional purposes such as wayfinding and locating places.

Widely received maps both create and draw upon established visual narratives, and these can be employed by authors in the framing of their own textual narratives. The styles and graphics of Ordnance Survey are a case in point, with regard to the British author Arthur Ransome (1884–1967), whose children's *Swallows and Amazons* stories are set in semi-fictional literary geographies of the English Lake District. Cooper and Priestnall, in their discussion of Ransome's landscapes, make the point that the Ordnance Survey's 1:25,000 maps of the area draw on extrinsic features of the landscape, which are in turn built upon from earlier mappings of the region. When the map is read by a map user, they construct a personal and internalized view of that geography that includes OS iconography and expression, drawing on any potential material which the user has previously been exposed to, which might inspire serendipitous associations with the landscape depicted. Cooper and Priestnall call this 'geo-specific intertextual context' (2011: 252). These associations thus feed forwards into human readings in, and experiences of, the same area, leading to 'reader-generated mappings' of landscapes, personalized traces which describe the reader's own pathway – whether literal or metaphorical – through the landscape, based on an intertextual reading of map, text, and natural environment (Cooper & Priestnall, 2011: 251–252).

It is also the case that the very act of constructing a narrative, of assuming the role of author/cartographer, creates an unequal power balance between the narrator and the narrated about. In the digital sphere, this is the core of Stephens's argument about the appropriation of web space by those who choose to participate in the creation of Web geography, and the exclusion of those who do not (Stephens, 2013; and see Chapter 5). All of these approaches rest on the still more fundamental assumption that the components of spatial narrative are necessarily relatable to abstract the quantitative geographies of spatial data, which – promulgated through the platforms and corporations of Silicon Valley – are, themselves, cultural constructs loaded with Western, technocratic values. The cultural implications of a perpetuation of the spatialized view of the world based on Cartesian digital geography on the *World Wide* Web, originally a product of Western technology, hardly needs elaborating (Mowery & Simcoe, 2002).

A key concern of the spatial humanities must surely be to deconstruct that assumption. 'Humanistic space', as opposed to abstract space, requires consideration of the cultural context in which a spatial narrative is formed, and this illustrates the limitations of spatial narrative derived narrative theory based on literary, Western traditions. Just as the linear, event-based 'beginning, middle and end' model does not work in the reading of spatial narratives on maps (these are what the reader makes them for any individual use of the map), so some non-Western contexts are far removed from structuralist narrative traditions in their conceptions of space.

One often-cited example of this concerns Aboriginal creation myths and artistic traditions. The ethnographies of the Aboriginal societies of Australia conceive of mythic creation traditions rooted in the movement of supernatural beings through and around the landscape. Topographic features such as mountains, streams, woods, sources of water and food – the fundamental elements of landscape that predicate survival – are formed by them as they engage with that landscape by moving, stopping, sitting down, hunting etc. Personal, individual experience of these features is irreducibly connected with these processes. Often, sites of significance in this narrative are marked by rock art, whose apparently stationary and immovable nature (a form of 'monumentality' according to Western traditions) belies the complex processes of movement and observation behind its creation (see Chapter 8).

Narratives of experience

The idea that a spatial narrative takes on its own cognitive being at the point where a reader/user/navigator makes use of an item of spatial discourse such as a map or a text is not new. Indeed theories of sensory, phenomenological engagement, and the creation of personalized internal narratives through the processes of physical experience with a landscape are at the core of post-processual archaeology, and many approaches to contemporary landscape studies. Tilley's attempt in *A Phenomenology of Landscape: Places, Paths and Monuments* (1994), which sought to re-think the development of prehistoric landscapes by interpreting them through contemporary embodied experience, is directly related to spatial narrative, and in particular the idea that spatial narrative must be able to cross media, different forms of written and imaged discourse, and the physical world. Most importantly for the conceptualisation of the 'deep map', Tilley concluded that '[a] critical understanding of spatial narrative requires that we investigate precisely why we prefer some plots or configuration of things rather than others' (Tilley, 1994: 32). He also stressed the importance of space as a subjective human construct: space that is formed by human intervention in the world and driven by human agency. By implication therefore he rejected the notion that these anthropogenic spaces are entities which can be understood through abstract space, disconnected from human agency, and representable in purely analytical and non-narrative terms.

Points of phenomenological engagement in the creation of narrative also occur around features of major focus in the landscape, such as monuments. These draw human attention, and act as foci for the enactment of shared experience within the landscape. Any large monument is likely embody a set of separations, all of which condition the delivery of spatial narrative: separations between form and function (the latter often very difficult to reconstruct in the cases of monuments of prehistoric periods, where there are no written records documenting contemporary usage and contemporary narratives), between performer and audience, and sometimes between whole communities.

Monuments are furthermore constructed with the whole idea of physical engagement first and foremost in mind – they are built to be experienced by those who see, hear and touch them, and participate in events which they inform and/or regulate.

Hadrian's Wall in Northumberland is an example of such a monument which combines multiple spatial narrative components, received both synchronously and asynchronously by multiple audiences (see Figure 6.1). The Wall, as at one time the northernmost border of the Roman Empire, has embodied many narratives of separation through time (Hingley, 2010), and – obviously – in space. Once again, we must look to the phenomenology of reception through intertextual relationships to understand how these narratives have changed over time, according to the social and cultural conditioning of those forming the narratives. In 1586 Camden noted of the Wall that 'It was called by ancient writers Vallum Barbaricum, Praetentura and Clusura by Dion. … Murus, Picts-Wall'; terms with undertones of militarism and separation. It does not take any great leap of the modern imagination to see such descriptions in the light of the border wars between England and Scotland which characterized the region of the sixteenth century in which Camden was writing.

Similarly, another antiquarian, John Collingwood Bruce (1805–1892), the Newcastle schoolmaster and inventor of the numbering system for Hadrian's Wall's milecastles and turrets which is still in use on today's maps, described it as 'The Roman Wall … a great fortification stretching from Wallsend on the

Figure 6.1 Hadrian's Wall
Photo: the author

Tyne to Bowness on the Solway' (Bruce, 1909: 15). Once again, it is hard to see this spatial narrative outside of the context of Imperial Britain, and the nineteenth-century establishment of the idea of the nation state. Writing a year earlier, T. Mommsen similarly channelled the *zeitgeist* of the European nation state, and contemporary tribal and national references, when he compared the Wall to the *limes* frontier lines of the Empire that stretched through Germania: 'They [The German *limes*] had not, like the Britannic Wall, the object of checking the invasion of the enemy … the Romans in Upper Germany did not confront their neighbours as they confronted the Highlanders of Britain, in whose presence the province was always in a state of siege.' (Mommsen 1909).

A similar critical understanding of time-critical contexts is needed to understand the Wall's spatial narrative in the present day. Nesbitt and Tolia-Kelly describe it as one shifting, through time, from one of the monument imposing a particular configuration upon the landscape, to one where the landscape – in the form of the close curation and management of English Heritage and the National Trust – imposes a configuration on the Wall in the present world (Nesbitt & Tolia-Kelly, 2009). They note that 'This approach seeks to enrich narratives of the Wall made intelligible through a body-centred account' (ibid.: 370). In this model, the essential aspect is the dynamic change of place, significance, viewpoint and human experience through time.

Conclusion

Any form of human discourse which describes or attests place has the potential to form a transmedial network that may become a spatial narrative. These structures (a) allow geographic information to be shared among multiple readers, and (b) have the capacity to encode subjective viewpoints of geography. Where they profess to encode 'ground truth', this is in fact an abstraction, usually stemming from Western frameworks such as the World Geodetic System. This does not in itself mean that digital environments, platforms and analytical structures such as GIS cannot be used to gain an understanding of spatial narrative, it simple means it must be recognized that they will form only *part* of a spatial narrative.

The critical tension between space which can be measured in extrinsic and geometrical terms and space which cannot brings us back to the question of how *both* types are received and understood by readers/audiences. This includes questions of how authorial (or cartographic) biases, whether unconscious or not, are dealt with. *Understanding* is the key: humanistic spatial information, the primary material of the spatial humanities is defined above as references to or descriptions of place in texts, manuscripts, old maps, material culture, and performative environments. Each is a medium of transmission from author/cartographer to reader/map user, and this requires us to consider how the author/artist/cartographer's understanding or perception of space aligns with – or challenges – the recipient's. For example Crane's map (above) contained a clear message about the British imperial project in 1886, but how was this received, both by Crane's contemporaries in Victorian society, and by colonial populations?

Spatial narratives as conceived by Tilley, Azaryahu and Foote etc. do not share the chronological/structuralist assumption of having a beginning, a middle and an end, as a more general understanding of narrative would have. Rather the structure is subjective and intrinsic to the reader. A spatial narrative is more akin to what Macfarlane describes as 'story maps', being 'how people understand themselves using landscape, by the topographies of self we carry within us, and by the maps we make with which to navigate these interior terrains' (Macfarlane, 2012: 26). This relates to Caquard's distinction between 'story maps', representing place as it is perceived by an individual or by a culture moving through it and a 'grid map', which is an authoritative abstract statement of things' location on the Earth's surface (Caquard, 2011). In other words, the latter are well supported by the kind of positivist geography familiar to millions from Google Earth (and other forms of internet mapping), which perpetuates a catch-all implementation of everything being represented in points, lines and polygons, with a definitive timestamp, whereas the former are not.

Ultimately, the construction of spatial narrative concerns the transfer of spatial information from a writerly context to a readerly one. Examples of it are encoded, what the processes of interpretation involved in that encoding are, and how it changes as a result of the transfer, are given here. What emerges most clearly from the discussion above is that spatial narrative is personal, often internalized, and as subjective as the human view of place itself. It accommodates positives and processual approaches, but is not constrained by them. It recognizes the importance of personal, phenomenological engagement. Spatial narrative is also fundamentally transmedial and intertextual. The construction of a personal spatial narrative, whether it is to explain, inform or coerce others requires connecting spatial cues in maps, texts, the real (physical) world and images. It is, in other words, very much akin to the principles, processes and practices of deep mapping laid out in Chapter 1.

A spatial narrative may be seen as spatial information that does not exist in isolation in any one medium or state. As noted in the introduction, the Internet and WWW has vastly expanded the walls of Plato's Cave, but to understand the spatial information and spatial data therein (and how this shapes human spatial identity), we must rethink the classical structure of the narrative of having a beginning, middle and end, and stages in between. Rather, in the Lefebvreian/Masseyian view of space, spatial narrative is constantly made and remade as it is told and re-told across media, as it jumps from one medium to another. As prisoners in the Cave, appreciating these distinctions is essential for any critical understanding of place in the human discourse, past or present.

References

Azaryahu, M. & Foote, K. E. (2008). Historical space as narrative medium: on the configuration of spatial narratives of time at historical sites. *GeoJournal*, 73(3), 179–194. Available online: https://doi.org/10.1007/s10708-008-9202-4

Barber, P. & Harper, T. (2010). *Magnificent maps: Power, propaganda and art*. British Library.

Biltcliffe, P. (2005). *Walter Crane and the Imperial Federation Map showing the extent of the British Empire (1886). Imago Mundi*, 57(1), 63–69. Available online: https://doi.org/10.1080/0308569042000289851

Bodenhamer, D. J., Corrigan, J., & Harris, T. M. (2015). *Deep maps and spatial narratives*. Indiana University Press.

Bruce, J. C. (1909). *The hand-book to the Roman Wall: A guide to tourists traversing the barrier of the lower isthmus* (6th ed.). Newcastle-Upon-Tyne: Andrew Reid and Co.

Brück, J. (2005). Experiencing the past? The development of a phenomenological archaeology in British prehistory. *Archaeological Dialogues*, 12(1), 45–72. Available at: https://doi.org/10.1017/S1380203805001583

Caquard, S. (2011). Cartography I: Mapping narrative cartography. *Progress in Human Geography*, 37(1), 135–144. Available at: https://doi.org/10.1177/0309132511423796

Collier, P. (2002). The impact on topographic mapping of developments in land and air survey: 1900–1939. *Cartography and Geographic Information Science*, 29(3), 155–174. Available at: https://doi.org/10.1559/152304002782008440

Cooper, D. & Priestnall, G. (2011). The processual intertextuality of literary cartographies: Critical and digital practices. *The Cartographic Journal*, 48(4), 250–262. Available at: https://doi.org/10.1179/1743277411Y.0000000025

Crampton, J. W. & Krygier, J. (2006). An introduction to critical cartography. *Acme*, 4(1), 11–33.

Fink, C. (2011). Mapping together : On collaborative implicit cartographies, their discourses and space construction. *Journal for Theoretical Cartography, 4*, 1–14.

Gosden, C. & Marshall, Y. (1999). The cultural biography of objects. *World Archaeology*, 31(2), 169–178. Available at: https://doi.org/10.1080/00438243.1999.9980439

Heffernan, J. A. W. (1991). Ekphrasis and representation. *New Literary History*, 22(2), Probings: Art, Criticism, Genre ((Spring, 1991), 1991)297–316.

Hingley, R. (2010). 'The most ancient Boundary between England and Scotland': Genealogies of the Roman Walls. *Classical Receptions Journal*, 2(1), 25–43. Available at: https://doi.org/10.1093/crj/clq001

Hodder, I. (1993). The narrative and rhetoric of material culture sequences. *World Archaeology*, 25(2), 268–282.

Leszczynski, A. (2009). Quantitative limits to qualitative engagements: GIS, its critics, and the philosophical divide. *The Professional Geographer*, 61(3), 350–365. Available at: https://doi.org/10.1080/00330120902932026

Lock, G. (2010). Representations of space and place in the humanities. In D. Bodenhamer, J. Corrigan, & T. M. Harris (eds.), *The spatial humanities: GIS and the future of humanities scholarship* (pp. 89–109). Indiana University Press.

Macfarlane, R. (2012). *The Old Ways: A Journey on Foot*. London: Penguin Books.

Mommsen, T. (1909). *The provinces of the Roman empire from Caesar to Diocletian*. Translated with the author's additions, by William P. Dickson. London: Macmillan and Co.

Moser, S. (2010). THE DEVIL IS IN THE DETAIL: Museum displays and the creation of knowledge. *Museum Anthropology*, 33(1), 22–32. Available at: https://doi.org/10.1111/j.1548-1379.2010.01072.x

Mowery, D. C. & Simcoe, T. (2002). Is the Internet a US invention?—an economic and technological history of computer networking. *Research Policy*, 31(8–9),1369–1387. Available at: https://doi.org/10.1016/S0048-7333(02)00069–0

Nesbitt, C. & Tolia-Kelly, D. (2009). Hadrian's Wall: Embodied archaeologies of the linear monument. *Journal of Social Archaeology*, 9(3), 368–390. Available at: https://doi.org/10.1177/1469605309338428

Pluciennik, M. (1999). Archaeological narratives and other ways of telling. *Current Anthropology*, 40(5), 653–678. Available at: http://www.ncbi.nlm.nih.gov/pubmed/10539944

Rowe, S. M., Wertsch, J. V., & Kosyaeva, T. Y. (2002). Linking little narratives to big ones: Narrative and public memory in history museums. *Culture & Psychology*, 8(1), 96–112. Available at: https://doi.org/10.1177/1354067X02008001621

Sontag, S. (1978). *On photography.* London: Penguin Books.

Stephens, M. (2013). Gender and the GeoWeb: divisions in the production of user-generated geographic information . *GeoJournal*78(6), 981–996.

Tilley, C. (1994). *A phenomenology of landscape: places, paths, and monuments.* Oxford: Berg.

Wright, J. H. (1906). The origin of Plato's Cave. *Harvard Studies in Classical Philology, 17*(May),131–142.

7 The structure of geodata

Introduction

Geographic data has a range of possible formal structures, which function on different levels. In Chapter 5, for example, we saw that crowdsourced academic knowledge, which is processed and regulated into academic narratives through professional regulation and peer review, adheres to the conventions of scholarly communication, familiar formal units such as the journal article and book; and is enabled by the involvement of professionally based academic participants (see also Hedges & Dunn, 2017). This is what distinguishes it from the kinds of unregulated and unreviewed and, largely, *unstructured* content which makes up open web geodata, such as that available via OpenStreetMap (even though individual place references therein also gain greater or lesser social traction through the visibility (hits, views, rankings etc.) that they do or do not acquire). This raises various questions relating to authority conferred by structure. This chapter explores how the process of regulating and structuring geographic data shapes the kinds of narratives and arguments it can form. Continuing the historical theme of this book, it also highlights that digitally structured geodata which forms part of scholarly narratives in the spatial humanities must be seen in the context of how principal ways of structuring it developed before the age of the WWW and Internet. As elsewhere, we will find that this means looking in to the longer prehistory of the spatial turn in the humanities.

Apart from the map or the atlas, the type of information structure most associated with geographical knowledge is the gazetteer. This is a term which usually denotes a list of places, usually accompanied by an absolute spatial reference such a latitude or longitude, a place name, or some other form of geometry. However, as in many other areas, the discursive and epistemological demands of the spatial humanities require questioning the structure(s) and concept(s) of the gazetteer in new ways. In many spatial research projects involving digital resources, the term 'gazetteer' itself, the kinds of authority a gazetteer confers, and the kinds of knowledge-creating epistemology it supports, has been reconsidered. Mostern and Johnson (2008) for example frame the creation of historical gazetteers in terms of mapping events rather than (just) places because – in agreement with the definitions of spatial narrative presented in this book – it is the 'event-ladenness' of places which make them significant for understanding place in the human record. The historical significance of Waterloo as a major battle in 1815, for example, is in most cases, likely to eclipse its significance as 'OpenStreetMap ID= 191673'; yet the latter might still be useful in disambiguating a textual attestation of 'Waterloo' from the train station in London. However, gazetteers remain the main vehicle for conveying and expressing formal spatial data and information, and for the socilaized construction of spatial narrative. It is thus worth spending some time examining the origins of the term and, by extension, the kinds of intellectual authority a gazetteer can convey.

Gazetteers share the authority and referencing structures of libraries, acting as curated points of entry for spatial information. Much 'authority' in the academic humanities begins with the processes by which researchers access information. It is a primary tenet of any university study skills programme, for example, that one should turn to sources whose authority can be verified, before turning to Wikipedia, if one turns to the latter at all. The management of any kind of scholarly communication thus always begins with some form of library, cataloguing or curation system, which (authoritatively) signposts the user towards verified sources of the information they are looking for. As in other areas however, the presence on the web of unregulated user-generated content has complicated scholars' relationships with these structures. For example Cole and Hackett note that using collaborative or crowdsourced methods to tag library catalogues for the purposes of searching and information retrieval disconnects communities from the 'gatekeepers of the cultural record', which undermines the very idea of the academic source (Cole & Hackett, 2010: 112–23).

A key problem for the spatial humanities in an academic context is that space and place, by their nature, resist the conventional constraints of librarianship – this is particularly true of the discursive and qualitative ways in which humanists place-make. Either spatial references are subsumed within the discourses of texts or other types of discourse; or they are contained within maps. Maps are similarly problematic to fit within wider conventional library structures, which are concerned with the ordering and storage of collections of text contained within books and manuscripts. They also resist those of art curation, which is concerned with the ordering and storage of art objects; and museum curatorship, which is concerned with the curation of objects.

Maps, of course, can be any of these things, and potentially combine the characteristics of all of their attributes. Library cataloguing systems were designed to catalogue books for the purposes of storage, search and retrieval. Both the book as a physical object, and fundamental elements of books' metadata (title, author, publisher etc.) lie at the core of library science as a discipline, and have done so for hundreds of years. However maps share neither the physical makeup nor the essential metadata elements of books, still less the places they represent. For this reason cataloguing them according to conventional library standards is highly problematic. As Ristow noted in the 1960s, 'maps persistently refuse to conform to established cataloguing and filing procedures' (Ristow, 1967). This means that the cataloguing systems of physical libraries served the needs of their users differently with regard to maps and books. Anyone who has attempted a search of the British Library's manuscript map collection, for example, will be aware of the limitations of having one's search limited to (modern) country, author/cartographer and a seemingly random date range.

Maps are difficult to classify in library terms. They can be classified according to geographical area, of course, but this only provides the most rudimentary search capabilities, and is even more limited in the case of pre-modern maps, which were produced before the advent of modern cartographic methods such as triangulation and, later, satellite imagery. They may thus relate only very approximately to the present-day layout of cardinal features such as rivers, lakes, coastlines etc.; and of course there is the certainty that the built environment will have changed over time as well. Within the gamut of the spatial humanities, the principal approach which those charged with preserving and facilitating access to map collections have rolled out is to use humans to georeference digital scans of old maps. In such scenarios, the user will examine the scan overlaid with a modern map, and identify 'control points' on both, where a feature recorded on the old map corresponds clearly to an identifiable feature on the modern map. A database of such equivalences can ensure that, when a user searches for a location with a modern latitude/longitude, any results from the collection of old maps will be retuned when they correspond with that modern location.

This tension gave rise to the (hugely successful) British Library Georeferencer project, whose purpose was to 'geo-enable' historic maps by asking participants to crowdsource ground truth points on scans of old maps, thus assigning geometrically accurate spatial coordinates for searching the collection and retrieving relevant scans. Once digitized and georeferenced, the maps can be viewed using online geographic technologies, and are geographically searchable due to the inclusion of latitude and longitude coordinates in the metadata. Some 725 maps were georefernced between 13 and 18 February 2012 by around 90 participants, to an excellent standard of data quality (Fleet, Kowal, & Pridal, 2012)

Many similar problems apply to the citation of cartographic material. The practice of citation in scholarly writing is inherently geared towards the book, the journal and the article, with less attention given to the citation of place. Some university libraries have produced guidelines on the citation of cartographic materials. For example, Harvard University's Library offers the following advice for the citation of printed maps:

> ***Printed map***
>
> *What should be included: In-text citation should contain, Map maker/author, Year of publication or issue in (round brackets).*

> *Reference List should contain, Map maker/author, Year of publication. Title of map. Map series, Sheet number, scale, Place of publication: Publisher.*
>
> (Source: http://rmit.libguides.com/harvardvisual/maps, last accessed 25 July 2018)

Similar examples can be culled from the websites of many other university libraries in North America and Europe; and many of these derive from the standard reference work on the subject, Kollen et al.'s *Cartographic Citations: A Style Guide* whose second edition was released by the American Library Association's Map and Geography Round Table in 2010 (Weimer, 2011). This guide is used in, for example, the recommendations of the Bibliographic Control Committee of the Association of Canadian Map Libraries and Archives in its guidance (Auringer Wood, 2012), whose 'basic form' for the citation of a map echoes the basic metadata elements of the book in the conventional physical library, of Author. Title. Edition. Scale. Place of publication: Publisher and Date.

From paper to digital citation

The rapid development of digital cartography since 2005 (the year of the production of Google Earth), and WWW cartography (see Chapter 2) highlighted the challenges that GIS and other spatial datasets presented for conventional map librarianship. This raises the further, and more complex, question of how one refers to a place in a (digital) document. In one sense, the emergence of the GeoWeb forces this question to be asked, by separating the products of cartography (maps) from the concept of the place as an informational object (whether qualitatively or quantitatively expressed); and thus undermining further the (already clunky) structures of library information systems' usage of metadata headings designed, conceptually, with (physical) books in mind. The Library and Information Science (LIS) community responded in 2008 when the ALA's Map & Geography Round Table (MAGERT) released a set of guidelines setting out the 'core competencies' in in GIS which map librarians should have in order to fulfil their professional duties (http://www.ala.org/magirt/sites/ala.org.magirt/files/content/publicationsab/MAGERTCoreComp2008.pdf, last accessed 26 July 2018). Kollen et al's guidance was refined further to take account of the new world of linked online geography. Brock University, for example, implements Kollen et al. for objects defined as 'Map created using GIS software':

> Map Title [map]. Scale. Data layers publisher and names [computer files]. Place of publication:
>
> Name of person who generated map, date. Using: computer software name [type of software].

> Version or edition. Software manufacturer location: Software manufacturer name, date of software
> release/copyright.
> (http://www.brocku.ca/maplibrary/guides/How-to-Reference.pdf, last accessed 26 July 2018).

Such approaches help to frame old maps within the context of contemporary scholarly authority by making them available with the critical apparatuses necessary for their use in modern research. However, the very process of associating individual places within a map, and equating specific points on the gridless manuscript map with the WGS84-grid on the modern one, highlights the fact that digital mapping, and digital cartography, generates *digital spatial objects.* The curatorial authority structures, in which Cole and Hackett characterize the role of librarians as the 'gatekeeper[s] of the cultural record' are thus replaced by other platforms. These gatekeepers are the keepers, developers and maintainers of digital gazetteers.

The gazetteer as an imperial tool

This represents a shift in curatorial emphasis from (physical) objects to digital ones, and this needs to be seen in the context of the developments described in Chapter 2, notably the emergence of Zook and Graham's 'DigiPlace' (Zook & Graham, 2007a); as well as in the models of academic knowledge co-creation discussed in Chapter 5, which rely on effective relationships between professional and non-professional communities. As the Internet and WWW grew and adapted to ever more widespread and accessible representations of place through the mid-2000s, so the humanities began to adapt their usage of internet infrastructures for the purposes of citation and referencing spatial data.

As noted above, gazetteers are lists of places, which often but not always attribute a name to each place, along with other attributes such as position, location, population, administrative significance etc. In doing so, they also enact an implied power relationship. Historically, this often manifested itself (as will be explored later in this chapter) as an expression of a powerful society exercising authority over a less powerful one. As structures of information and knowing, gazetteers define distinctions between the knowing and the known-about, in much the same way as maps do. They had – and in some cases have – the power to enact precisely the kinds of power imbalances seen in the exploration of neogeography, but on a formal and institutional scale. Many early-modern gazetteers contain the same pretensions to encyclopaedic, all-knowing knowledge about 'other' places as the chorographic works discussed in Chapter 2, and are clear about their quasi-imperial context. For example, the 1762 work titled

> *The American Gazetteer: containing a distinct account of all the parts of the new world, their situation, climate, soil, produce, former and present*

> *condition, commodities, manufactures and commerce; together with an accurate account of the cities, towns, ports, bays, rivers, lakes, mountains, passes and fortifications; the whole intended to exhibit the present state of things in that part of the globe, and the views and interests of the several powers who have possessions in America.*

… represents a view of the process of landscape recording with a firm rooting in the context of 'powers who have possession', and is somewhat reminiscent of the chorographic texts discussed in Chapter 2.

The imperial roots of the gazetteer as a means of documenting the literal and intellectual ownership of an area come together very clearly in Walter Crane's *Imperial Federation Map Showing the Extent of the British Empire* of 1886. In this richly illustrated map, Crane depicts the extent of the British Empire surrounded by stylized caricatures of its residents. These include eroticized and orientalized female figures on the right-hand side, native figures down the left, and the figure of Britannia herself in the bottom centre. Framed by these, the map itself contains seven tables, placed so as to overlap with key trading routes, documenting the area, volume of trade and population of each colonized state in 1851 and 1866. The clear link between the quantifiable attributes of a set of places, as seen by the governmental authority of another in both these early examples, exemplifies precisely the partial nature of the gazetteer structure for much of cartographic history before the twentieth century.

The first use of the word 'gazetteer' in its more neutral meaning as lists of places for reference purposes is likely to be the title of *The gazetteer's: or newsman's interpreter. Being a geographical index of all the considerable cities, patriarchships in Europe,* by Lawrence Eachard (c. 1670–1730). This volume explicitly sets out to list place names and their meta-information including population, administrative area and hierarchy, and the spellings of the names insofar as these are known systematically; but, echoing the constraints acknowledged by the chorographic antiquarians discussed in Chapter 2, notes that there are practical limits to the scale at which it is possible to represent places in such a format – '[t]he Reader is not expected to find every little place that is sometimes mentioned in the Gazettes and News-books, for that would be a work of vast Difficulty, and little use' (Eachard, 1704). The limitations of the paper medium again regulated the scale and granularity of the information that could be published.

Gazetteers and the WWW

Gazetteers of the Web era are not so constrained. These lists of spatial data differ from the wild west of 'DigiPlace', in that they are formal structures of place and its meta-information, rather than being a '[s]ubjectively designed interactive space that influences how people interact with their local environment' (Zook & Graham, 2007a: 480). Online, they provide the capacity for *referencing* places which might otherwise be ambiguous, fuzzy, contested or otherwise unclear – a means for referencing ontologically discrete locations, through hyperlinking, in digital documents.

One of the best-known WWW gazetteers is GeoNames.org. GeoNames is a crowdsourced gazetteer containing some 10 million geographical names, categorized according to 645 feature codes (information from www.geonames.org, retrieved 14 June 2018). This resource allows webpages and any data therein – such as Wikimedia pages referring to particular places, whether named or not – to be georeferenced. The Wikipedia entry for 'Knossos', for example (https://en.wikipedia.org/wiki/Knossos), contains a link to the 'geohack' section of the Wikimedia site, providing cross-references to some 21 web gazetteer services, of which GeoNames is one. This enables the page to refer unambiguously to one place, separating it ontologically as well as geographically from all others. In other words, this may be seen as a *citation* of these gazetteers – the page draws on the referential authority of the GeoNames entry for Knossos at http://www.geonames.org/259627. While this contains a latitude/longitude as an attribute, the primary referent is the numerical Universal Resource Indicator (URI), and a unique number string it contains. This allows automated disambiguation between different places with the same name, such as Birmingham, Alabama (http://www.geonames.org/4049979) and Birmingham, England (http://www.geonames.org/2655603); or, as in the case cited above, the battlefield of Waterloo and the station that was named in honour of it. One limitation with GeoNames for use in spatial humanities research is that relevant Feature Types, such as the 'archaeological place/prehistoric site' is relatively underutilized and/or inconsistently applied. For example, the major Neolithic monument in southern England, Stonehenge, is associated with this category (http://www.geonames.org/2636812); whereas the nearby Neolithic site of Avebury is not (http://www.geonames.org/2656760), the latter being simply listed as a 'third order administrative division', a reference to the town of Avebury. It may therefore be said that GeoNames is a critical part of the GeoWeb which supports disambiguation, linking and (curated) geoparsing output, but as a (spatial) humanities research tool it suffers from the same kind of limitations as OpenStreetMap, discussed in Chapter 5 – it is a core database, rather than a set of services which researchers can make use of, or at least which can be easily used for the purposes of research.

One such example of 'under the hood' digital research which web gazetteers *can* be used for is supporting tokenizing algorithms to parse electronic texts and automatically identify places named therein, a process known as geoparsing (Grover et al., 2010). Most of the digital products and platforms referred to up until now rely on some form of digital gazetteer, and these underlie various forms of spatial/digital humanities web resources (Luchetta, 2017). One well-known example of a tool which relies on geoparsing from multiple gazetteers is *Recogito*, a service produced by the Pelagios project, which uses an external geoparsing service (the Edinburgh geoparser) to automatically analyse digital texts and identify place names, displaying corresponding entries from multiple gazetteers (Simon, Barker, Isaksen, & de Soto Canamares, 2015). The user – with the benefit of human expertise – can then edit the output to correct for any errors the machine has made.

One project which exploits this functionality for the spatial humanities research into the ancient world is the Pleiades gazetteer (https://pleiades.stoa.org; see Elliott & Gillies, 2009). Pleiades, which contains some 36,000 places, provides a model of community-curated production of URIs for ancient place references. This allows the complexity of physical, geographical and historical connections, which may change over time, to be expressed as data structures. For example the Santorini Archipelago in the Aegean Sea is expressed as a single URI (https://pleiades.stoa.org/places/808255902); and this 'place' contains as sub-units the islands that make up the archipelago – Thera, Therasia and Aspronisi, as well as the volcanic island masses in the middle of the Santorini caldera, Paleo Kameini and Nea Kameini. Each have their own URIs, which have 'child' relationships to the archipelago URI. However, it also contains a third caldera island, 'Mikra Kameini'. Mikra Kameini was surveyed in 1849 by *HMS Volgae*, under the command of Thomas Graves (Leycester, 1850). However, it is not extant today, as it was joined to the larger nearby island of Paleo Kameini, probably as a result of volcanic activity in the caldera during 1866–70. The Pleiades gazetteer structure allows this otherwise invisible nuance of the Santorini archipelago's history to be captured and fully referenced. As well as capturing the parent and child relationships between places, and abstractly defining the places themselves, every Pleiades URI ascribes authorship of the person responsible for creating it.[1]

This requirement is in contrast to the analytical functions of GIS; indeed the Pleiades team once described the project as an 'un-GIS for ancient geography' (http://dh2011abstracts.stanford.edu/xtf/view?docId=tei/ab-192.xml;query=;brand=default). Pleiades began as an effort to digitize the data underlying the *Barrington Atlas of the Ancient World*, and open up the resulting dataset to refinement and expansion by the research community (Elliott & Gillies, 2009). This is a strong illustration of the key distinction made above between mechanisms for storing, retrieving and citing major reference works such as the Barrington, and accessing and citing the structured data it contains.

In this project, every name of every place is represented by a URI, which provides an authoritative and exclusive web citation for that place, which can be used and re-used in any web mounted media. This is an approach which

1 The emergence of Pleiades as a key resource reflects the fact that Ancient World geography is a field heavily dominated by text, and this means it requires secure, authoritative structures for referring to place, and navigability of these is a key requirement for scholars. One of the author's earliest memories as a PhD student in the early 2000s was a senior academic, having just purchased the Barrington Atlas, raging that they had 'just spent hundreds of pounds on this Atlas and it doesn't have an index!' (it turned out the index was in a later volume).

blends web cartography, or at least the underlying logic of web cartography, with Semantic Web principles and information science. While relatively simple in concept, it has paved the way for many important initiatives, most notably Pelagios (http://commons.pelagios.org). The purpose of Pelagios is to provide a mechanism for cross-searching distributed gazetteers, by linking them together at the metadata level using a common Resource Description Framework (RDF) protocol, the *Pelagios Gazetteer Interconnection Format* (https://github.com/pelagios/ pelagios-cookbook/wiki/Pelagios-Gazetteer-Interconnection-Format) (Simon, Barker, Isaksen, & de Soto Canamares, 2015: 53). This allows the same places represented in different gazetteer collections using their own local URI formats to be aligned, associated and connected. Such cross-collecting links of common places can be valuable for searching and retrieving information online from trusted sources. For example, Pelagios underpins a side window on the Pleiades page for each place record giving entries in third party collections which mention that place. For example, the Pleiades entry for the Pan-Hellenic shrine at Delphi (http://pleiades.stoa.org/places/540726) lists 'Related Content from Pelagios' which includes 56 'federated' records from four different sources, including coin entries from the database of the American Numismatic Society. Pleiades and Pelagios also inspired the Heritage Gazetteer of Cyprus, a project co-led by the author (http://cyprusgazetteer.org), which sought to extend the principle of relating URIs to places attested in sources, but added a distinction between 'single footprint' places (Archaeological Entities in the language of the gazetteer), and those with wider and less clear extents. It also allowed separate associated URIs for the same place referred to in different sources and with different toponymic spelling.

The WWW gazetteer provides a means of 'citing' place, as represented by a numerical string rather than any of the traditional geographical affordances of coordinates, place names, descriptions etc. It also stands in contrast to the cumbersome referencing structures that went before the digital age as described above. A scholar or author can refer to any unique place in a gazetteer, usually in the form of an embedded hyperlink, or as underlying code. But just as the structures of the library presented problems for the organization of spatial information in physical objects, the referencing of digital place in gazetteers presents a new set of challenges, which require revisiting the colonial and postcolonial connotations of the first gazetteers, described briefly above.

Case study: Cyprus

The process of creating gazetteer entries – i.e. the process of defining places in the first instance, and describing the various parent/child relationships between them, and the assigning of entries to feature or gazetteer categories – can be problematic and, contrary to the binary impression given by geoparsing applications, is inevitably subjective. In the spatial humanities, it can never be a purely machine effort; human intervention and expertise is always required.

The often complex relationship between quantitative and qualitative place in text is discussed in more detail in Chapter 4; but inevitably, when a list of places is extracted from a text, one is creating a gazetteer-like information structure, and in the process encoding textual bias. As noted above, the origins of the gazetteer as a format lie, by and large, in quasi-imperialistic description of one encultured space by the inhabitants of another; and while the post-Enlightenment age (in particular) may have ushered in a form of the genre of gazetteer which appears more list-like and neutral, like any other type of information structure, these are prone to reflect describer/described, mapper/mapped power structures.

Such structures are particularly significant when dealing with regions that are highly contested by different political or cultural authorities. An effective digital gazetteer structure which represents the etymological and geographical variation of the same places described in different accounts, and which accounts for different political social and legal views of place in the present day, can be extremely complex. A place which exemplifies this is the island of Cyprus.

Cyprus stands in contrast to the kind of region which major WWW gazetteers such as Pleaides were designed to represent. Rather than being a very large geographical area with a large number of places described in relatively limited detail, Cyprus is a small area whose places have a very deep and rich history, spanning the Greek, Turkish and English languages, amongst others. It has always been a profoundly contested region, having always resided at the 'crossroads of civilizations' of Europe, Africa and Asia. It is a particularly interesting case study for exploring the complexities of contestation in geographic and gazetteer data structures, because in the late nineteenth century, this complexity was recorded and documented in a short space of time using modern, and Western geographic data standards.

In 1878, Great Britain took over the administration of Cyprus from the Ottoman Empire, after some three centuries of Turkish imperial governance. In a passage which perfectly encapsulates the imperialistic nature of detached narrative description of one culture by another, H. Johnstone, R.E., who was stationed in Cyprus at the time, wrote in 1881 of this handover:

> During the three hundred years which have passed away since these stirring times [the introduction of British rule], the Island sank into insignificance … until one morning in July just three hundred years after Lala Mustapha had landed his legions in Limassol Bay, the inhabitants of the island awoke to find the British flag hoisted on the old battlements of their Capital, and Ottoman misrule forever at an end: it suddenly became the focus of interest of all Europe, the records of its past history were easily sought out and read, and new books were written to tell of its ancient splendour and prosperity, (Johnstone 1885: 59)

One key requirement of the new authorities on the island was an authoritative map, not only to determine the main strategic points on the island from a military point of view, but also to arrange for the collection of tax and legal administration. The officer tasked to lead the production of this map was Herbert Horatio Kitchener, then a lieutenant in the Royal Engineers, subsequently the commander of British forces in World War One. Work on a comprehensive trigonometrical survey of the island began in 1879, under Kitchener's direction. At the time, there seem to have been some disagreements amongst the British authorities as to the map's true purpose, the colonial administration wishing to establish robust systems for raising revenues, but with Kitchener himself aspiring to produce a scientifically accurate account of the island's topography, employing the latest military and cartographic techniques (Given, 2002: 10). Kitchener himself seems to have become frustrated at his posting away from the front line, writing to Sir Robert Biddulph, the British commander in Cyprus in August 1882:

> I can't help feeling that my remaining here in a civil capacity while military service was offered might be used against me in my future career. My greatest ambition up to the present has been to finish the map of Cyprus, and there is nothing I should so fear as not being able to do so after these years' work, but at the same time I feel sure that you will agree that a soldier's first duty is to serve his country in the field when an opportunity is offered him, and not to remain at his desk while others are fighting.'
> (Unpublished letter, Kitchener to Major General Biddulph, UK National Archives reference PRO 30/57/1/5)

However, Kitchener's Map, published by Stanford's of London in 1885, was based on the first island-wide trigonometrical survey, in 15 bound sheets, at a scale of one inch to one mile. The immediate effect of this map was to impose an abstract – and highly colonialized – geographical data structure on the island for the first time, replacing the decentralized and distinctly informal structures that had gone before, characterized by Johnstone as 'Ottoman misrule'. Kitchener's approach to place names was hampered by the fact that then, as now, individual settlements were frequently referred to by the habitants by names in both Greek and Turkish, with great local variation in style, attestation and linguistic nuance. Despite this, as many names as possible were gathered by surveying parties, which visited villages and settlements, questioning members of the local communities about the place names, their dialect, pronunciation and history of the settlement's name (Shirley, 2001). On the map itself, these names were transliterated into English. Sheet 15 of the map provides a letter by letter guide to the transliterations used.

Kitchener's Map provides an excellent case study for the creation of a formal geographical data structure for a particular area. It breaks down the features of the island's landscape into ontologically definable parts, some of which reflect the formal cartographic processes of the British army at the time; and some reflecting the characteristics of a census (Britain undertook a formal census of its new possession in 1881, which Kitchener seems to have drawn upon) – for example Muslim and Christian settlements are marked with respectively a crescent and a cross. This conjunction of imperial ambition and scientific cartography is representative of the appropriation of local geographies into data structures which might outwardly profess scientific neutrality, but which of course are anything but. As the archaeologist Michael Given has written of this particular example: 'By analyzing processes of demarcation in colonial Cyprus, we are also investigating forms of colonial knowledge and methods of control. It was not just fields and forests which were forced into a closed classificatory system, but the people as well' (Given, 2002: 2).

Geographic structures have continued to shape political and cultural aspects of life in Cyprus in its more recent history. After a tumultuous near-century, Cyprus gained independence from Britain in 1960; and 14 years later, Turkey occupied the north part of the island, partitioning it in its present form, with the so-called 'Green Line' separating Turkish Cyprus in the North, from the Greek Cypriot Republic of Cyprus in the South. As part of its peacekeeping efforts following 1974, the United Nations published a *Complete Gazetteer of Cyprus* (1987). Employing a methodology much similar to that of Kitchener's fieldwork expeditions to interview local inhabitants, this document seeks to provide a comprehensive and authoritative version of each place name in Turkish, English and Greek (Christodoulou & K⊠nstantinid⊠s, 1987). Such authorities allow local administrations to negotiate the difficult relationships between 'official' and 'unofficial' place names in administrative documents, and they also allow researchers to undertake spatial humanities projects, and build spatial humanities resources. One such resource seeks both to aggregate data from different resources using a gazetteer framework as the primary referencer, but also to express *at the same time* in a more nuanced way how historical place, as attested through primary textual and cartographic sources, can be explored. This project is the Heritage Gazetteer of Cyprus (HGC).

The HGC draws on the principles and practices of the Pleiades gazetteer (see above). It allows scholars working with historic texts or manuscripts to mint a canonical URI for any place name, which is stored in the database as an *attestation*. As noted above, the complexity and density of place attestations in Cyprus means that it represents a 'thick' layer of data in a limited area, rather than a 'thin' layer across an unconstrained region (such as the Greek or Roman orbits). This means that when minting a set of HGC URIs through the process of close reading, as opposed to geoparsing, the user is encoding a set of expert judgements in that URI.

The core requirement of any new entry in the HGC is primarily that it should have a name. That name must be supported by an authoritative and referenceable source. It is up to the reader/user to exercise their judgement (which can be geographical, linguistic or even philological, in the case of manuscripts) in how the attestation is to be transcribed, and how the record is to be georeferenced – i.e. associated with geometry on the ground.

A key problem when designing the HGC was working out how best to support and express those judgements. A close reading of an initial set of texts identified spatial footprinting as a major problem. In a large gazetteer at the scale of the world, or of a large region (such as the Greco-Roman world), a point with a latitude and longitude is generally sufficient for the purposes of most researchers. There are some exceptions – for example, Pleiades provides fine-grained spatial foot printing for some key sites such as Aphrodisias in Anatolia, the site of intensive research, transcription and editing of epigraphic inscriptions, which draws on the Pleaides gazetteer for geo-references (Bodard, 2008). As intimated above, the WWW is a format and vehicle that is amenable to gazetteers on this scale. But if one is comparing place attestations across closely read texts then questions of judgement with regard to spatial foot printing arise. For example, when Jean Richard refers to 'Baffe' in 1423, how does the spatial footprint of the place he is referring to relate to the 'New Paphos' referred to by D. G. Hogarth in 1889? Both refer to the same town, but philosophically and literarily they might rest on entirely different assumptions about its extent, limits and spatial ontology and its 'trans-cultural themes', in the sense described by Ballatore (2016: 12). Similarly, if one wishes to encode, and generate a URI for, the Bronze Age site of Palea Paphos, one is confronted by the fact that any contemporary name attestation for the site is lost because it existed before any written records, and also that much of it may be unexcavated. This compounds the problem of where one defines its boundary in a way that makes sense as the coordinate reference in a gazetteer that conforms to modern WWW geo-standards. Any literary attestation that does give a name for the site is a modern construct, which will likely associate the site with the (modern) name of the nearest (modern) settlement – in the case of Palea Paphos, the village of Kouklia. Another important factor is the time at which the attestation was made, and the time to which it refers. Nea Paphos, to continue with the example, represents a 're-founding' of the old in the fourth century BC as a new costal settlement, so 'Paphos' as a gazetteer concept could in fact refer to either, but is qualified by the point in time at which the attestation is in use.

The HGC addresses this problem in a high-level manner by distinguishing between 'Archaeological Entities' (AEs), which are units with their own discreet footprint, which were thus logically constructed within a discrete phase or time frame; and 'Historical Units', (HUs), which are regions or areas with no such discernible or identifiable footprint (see Dunn, 2017 for discsussion). This draws to some extent on the theory of the single context recording system in archaeology, where a context, such as a fill, a cutting, a post hole etc. – which can be just as objective to define as places in texts – is treated as an independent, yet linkable entity with its own unique identifier, which forms part of larger a matrix of relationships with other contexts. This analogy works, insofar as it acknowledges the uniqueness, and thus the identifiability, of any one place. However, the definition of the polygon is, of course, a subjective and discursive process, depending simultaneously on the reader's interpretation of the text, and their interpretation of the landscape, as relayed either through a Google Maps satellite image, or map tiles drawn for the Ancient world Mapping Center (see http://awmc.unc.edu/wordpress). The 'bird's eye' definition of the attested place highlights a philosophical tension between edges created conceptually by digital tools, and place defined though phenomenological engagement (or indeed the kinds of narrative of experience described in Chapter 6). Casey, for example, in asking the question 'do places have edges', proposes a distinction between edges which provide 'limits from which' and 'limits towards which' (*terminus a quo* and *terminus ad quem*); a kind of 'centre versus periphery' approach to boundaries, which might or might not be permeable (or indeed physical) (Casey, 2011). Casey proposes 'two genres of edge', those of *sites*, which have borders, and those of *places* which have boundaries (Casey, 2011: 72). The former is determinate and contains, the latter is osmotic and indeterminate. Casey contextualizes this with a consideration of the edges of *events* – that is to say the characteristic of events, their 'capacity to harbour surprise, their spatio-temporal character, above all their pivotal role in the change of one place into another … what holds for the edges of events holds for the edges of places' (Casey, 2011: 70). These are the kinds of very subjective judgements HGC users must make when defining the geometries of HUs – e.g. the extent to which an attestation represents a defined place, a geometry enclosing a region discursively referred to, or an event – or all three.

Every entry also requires the user to supply two dates, of the time period when the attestation was in use as described above, and for the time at which the attestation was published. This acknowledges the temporal, as well as the spatial, granularity of the places attested. It also underlines the way in which the HGC combines bibliographic practices (such as dating a reference) with digital referencing systems. In the HGC, a URI can be minted for any AE or HU. Dealing with dates in this manner also allows the process of naming a place to be treated as an *event*, in the manner described by Mostern and Johnson, who state that 'the existence and attributes of named places are historical events' (Mostern & Johnson, 2008: 1105). Mostern and Johnson furthermore make the important point that privileging the details of historical narrative over place in gazetteers is essential for the understanding of historical processes. The HGC adopts such a position, with the geometry forming only one aspect of the information set for each entry.

In the case of AEs, a point geometry is assigned within the spatial footprint of the object and, because this will necessarily be a unique physical place, there is also the capacity to add a GeoNames URI, which (potentially) enables the gazetteer to be exposed to other gazetteers which incorporate GeoNames references, although at the time of writing, this capability has not been utilized. This allows the references for AEs to be referenced in any third-party digital publication – in the case of the HGC project, this was utilized in the Inventory of Byzantine Churches on Cyprus (Papacostas, 2013). The use of the HGC in this context suggest a blurring of the line between conventional referencing of spatial information in the conventional library structures referenced above, and digital, URI-based ways of referencing more abstract place and spatial information.

HUs are more problematic to define as spatial footprints. Also, the decision as to whether a particular place should be represented as an AE or and HU is frequently one which relies on academic judgement. For example, Venetian Nicosia is undoubtedly a 'place' which developed in a composite manner over a period of time, yet spatially it is easily defined by the extant city walls. In the HGC, the geometry of an HU is defined by a polygon, which is subjectively defined by the user accessinga map interface. In most cases, this will be an interpretive and subjective approximation of the place the user wishes to describe and represent, and will thus bear more resemblance to the conceptual, or 'readerly' in the terminology of Cooper and Priestnall (2011) rather than geographical reality.

Every HU established must have such a geometry, but (as with AEs) it may have any number of different attestations associated with it, and these can be added by multiple users. The constraint is that each entry must be supported by a publicly available reference, or a canonical URI. This means that both references in texts, and place names described on maps can be treated as being associated with the same HU. The Nea Paphos example employed above contains the attestation *Paphonea*, a Venetian interpretation of the name used by the cartographer Pierre Moulart-Sanson in his 1720 map, the *L'Isle (et autrefois royaume) de Chypre*, with a canonical URI reference pointing to a high-resolution digitized scan on the map, which appears on the web site of the Sylvia Ioannou Foundation.

The Heritage Gazetteer of Cyprus therefore combines several elements of representing and referencing humanistic place. There is a conventional bibliographic reference, a URI, subjective formulation of place on the screen by close reading, and the combination of different media types, such as text (of any genre) and maps. It might therefore be said to combine some of the characteristics of a deep map. It encourages one to approach various types of humanities content with a critical approach to how it represents place, and then encourages the reader to consider how that place is best represented in the context of that material. It also allows the overlaying of different datasets (such as attentions of place in text of different genres and on maps). At the same time, it enables computable attestations of places to be linked to other datasets and other gazetteers online. This has been achieved by federating the HGC into the Pelagios federation of online gazetteers.

It is also possible to compare the subjective 'container' boundaries of HUs with the landscape at various points in history. Recently, it became possible to overlay HU polygons with a high-resolution scan of the Kitchener Map. This allows one to compare how subjective readers have defined places, versus the scientific, imperial-era mapping of Kitchener. This shows, for example, how the HU polygon for Famagusta differs from the footprint mapped by Kitchener – the former being a trigonometrically surveyed overview of the actual 'border' of the place (in Casey's term), versus the record creator's subjective 'boundary-like' view of the same site. There are 17 different attestations which reflect different spellings of the site in different time periods, in some cases recorded long after that period of usage. The HU of Famagusta is thus an imperfect, yet multivariate and multivocal representation, which reflects the historic 'depth' of the place.

Information treated in this way can also be used in more conventional GIS analyses, which can assist in reading HGC texts in new ways. For example, George Jeffrey's *Historic Monuments of Cyprus* of 1918 is an encyclopaedic account of the island's archaeological and architectural features. A filtered search on HGC entries can allow one to visualize all the HUs and AEs attested by Jeffrey as an overlay. Also, exploring the AE geometries means they can be subjected to various forms of analysis. For example, he describes eight locations he visited on a journey from Perakhorio to Nicosia. These can be exported into a GIS and plotted as points. If they are overlaid on to a Digital Elevation Model (DEM), then it is possible to render a viewshed analysis over them. This calculates which pixels in the DEM – i.e. which 30x30 meter squares of the landscape – can be seen from each point. This allows us to see that in this particular passage, the majority of the viewsheds extend east of the points, with mountainous regions to the west constraining the views, with the exception of Kidhonia and Lazania, which are monastery sites. This provides an additional visual context with which to read the passage in question.

Conclusion

The emergence of the WWW-readable gazetteer highlights, or rather reiterates the problem highlighted in Chapter 4 in relation to Conan Doyle's London, which is how one represents, on a map, the actual spatial footprint of a geo-database record. The WWW is very good at actualizing place in binary terms, but less good at representing its human, cultural and social attributes. It also requires spatial footprints to be defined in a highly empirical and positivist manner. The HGC did not make any pretence at solving the problem of defining boundaries with the AE/HU distinction, but the computational representation of 'inhabited' versus 'observed' space online, or in a GIS, remains. How does one define boundaries when mapping cultural, historical or archaeological sites, especially when using computers? (Lock, 2010; Ballatore, 2016).

Others working the realms of the spatial humanities and digital gazetteers have highlighted the 'eventfulness' of place, and the interchangeability of digital representations of the two (e.g. Mostern & Johnson, 2008). Transferring this logical distinction to the digital realm however means the imposition of binary (i.e. database-like) boundaries to both events and places. However, as with Pleiades and Pelagios, the HGC's map on the website is *not* its key output. All these projects are concerned not with the representational expression of humanities data, but rather with a post-representational understanding of it, enabled by digital data structures. The maps themselves are, at most, WWW mounted visual search tools drawing on existing web standards, which conform to the binary data protocols of the web and serve as a visual medium for an interpretation of the spatial data, as represented in coordinate form; a realistic accommodation of the Cartesian reality of the Internet.

A key aspect of research in the digital humanities is the structure and format of information. Indeed, it might be said that the evolution of the field from 'Humanities Computing' to 'Digital Humanities' (see Svensson, 2013) represents a shift of emphasis from the creation of such structures in digital environments, to the use of those structures in the creation of knowledge and understanding about the human past. Gazetteers have always been descriptive artefacts which, conceptually, provide a place with a definition according to set standards. As such, they have always had a certain agency of power, in the same way OpenStreetMap has an agency of power for some communities at the expense of that of others. In their earliest instances, they are explicitly colonial projects, providing vehicles for the description of the occupied by the occupiers. Kitchener's 'gazetteer-like' map of Cyprus is a good example of this; and the contrast with the Ottoman styles of governance which went before (and the way these are described by the new British authorities) enforces the notion that the Cartesian or binary modelling of place channels a particularly Western brand of colonialism.

In the digital age, gazetteers, much like digital maps, act as both a restrictor and a stimulator of spatial narrative (Caquard, 2011). They restrict narratives by virtue of the formal structure they impose through the absolute necessity for any Web gazetteer entry to conform to WWW data standards in order to serve the basic purpose of *any* Web data, which is to be connectable to other datasets with the same information standards. But they stimulate such narratives by enabling the creation of federated, non-centralized networks of such references. The Heritage Gazetteer of Cyprus, whose development is outlined above, seeks to enhance the stimulative aspects of the gazetteer format (by providing a unique web-readable ID of every place) and minimize the restrictive elements by allowing the user an imperfect, yet transparent, means of defining the place geometric reference. Most importantly however, it combines the capacity to define place in this relatively discursive way with one of the most familiar and established means of ensuring academic traceability and authority, the bibliography.

References

Auringer Wood, A. (2012). ACMLA Recommended best in Citation of Cartographic Materials Compiled by Alberta Auringer Wood Bibliographic Control Committee of the Association of Canadian Map Libraries and Archives (ACMLA) Basic Form Examples Map on a single sheet. Bibliographic Control Committee of the Association of Canadian Map Libraries and Archives (ACMLA). Available at: https://acmla-acacc.ca/docs/ACMLA_BestPracticesCitations.pdf

Ballatore, A. (2016). Prolegomena for an ontology of place. In H. Onsrud & W. Khun (eds.), *Advancing Geographic Information Science: the past and next twenty years*, 91–103. GSDI Association Press.

Bodard, G. (2008). The inscriptions of Aphrodisias as electronic publication: A user's perspective and a proposed paradigm. *Digital Medievalist, 4* . Online only – available at https://journal.digitalmedievalist.org/articles/10.16995/dm.19/

Caquard, S. (2011). Cartography I: Mapping narrative cartography. *Progress in Human Geography*, 37(1), 135–144.

Casey, E. (2011). Do places have edges? A geophilosophical inquiry. In S. Daniels, D. DeLyser, J. N. Entrikin, & D. Richardson (eds.), *Envisioning landscapes, making worlds: geography and the humanities*, 65–73.

Christodoulou, M. & Kōnstantinidēs, K. (1987). *A complete gazetteer of Cyprus.* Cyprus Research Centre: Department of Lands and Surveys.

Cole, R. & Hackett, C. (2010). Search vs. research: Full-text repositories, granularity and the concept of 'source' in the digital environment. In C. Avery & M. Holmlund (eds.), *Better off forgetting? Essays on archives, public policy and collective memory*, 112–123. Toronto.

Cooper, D. & Priestnall, G. (2011). The processual intertextuality of literary cartographies: Critical and digital oractices. *The Cartographic Journal,* 48(4), 250–262.

Dunn, S. (2017). Praxes of 'The Human' and 'The Digital': Spatial Humanities and the digitization of place. *GeoHumanities*, *3*(1), 88–107.

Eachard, L. (1704). *The gazetteer's, or, newsman's interpreter : Being a geographical index of all the considerable provinces, cities. Patriarchships, Bishopricks, Universities, Dukedoms, Earldoms, and such like ; Imperial and Hanse-Towns, Ports, Forts, Castles, &c. in Europe.* London: John Nicholson and Samuel Ballard at the Kings-Arms in Little-Britain.

Elliott, T. & Gillies, S. (2009). Digital geography and classics. *Digital Humanities Quarterly*, 3(1). Available at: http://www.digitalhumanities.org/dhq/vol/3/1/000031/000031.html

Fleet, C., Kowal, K. C., & Pridal, P. (2012). Georeferencer: Crowdsourced georeferencing for map library collections. *D-Lib Magazine*, 18(11/12), Available at: http://www.dlib.org/dlib/november12/fleet/11fleet.html

Given, M. (2002). Maps, fields, and boundary cairns: Demarcation and resistance in colonial Cyprus . *International Journal of Historical Archeology*, 6(1), 1–22.

Grover, C., Tobin, R., Byrne, K., Woollard, M., Reid, J., Dunn, S., & Ball, J. (2010). Use of the Edinburgh geoparser for georeferencing digitized historical collections . *Philosophical Transactions of the Royal Society A: Mathematical, Physical and Engineering Sciences*, 368(1925).

Hedges, M. & Dunn, S. (2017). *Academic crowdsourcing in the humanities: Crowds, communities and co-production.* Chandos Information Professional Series

Johnstone, H. (1885)*The Birth Of Cyprus.* By Lieut. H. M. Johnstone R. E. Dedicated By Permission To H. E. Sir Robert Biddulph, K.C.M.G., C.B., High Commissioner of Cyprus. — Records Of The Ottoman Conquest Of Cyprus. By Major D. A. Donne Royal Sussex Regiment. — Cyprus Guide And Directory 1st April 1885.

Leycester, E. M. (1850). Some Account of the Volcanic Group of Santorin or Thera, once Called Calliste, or the Most Beautiful. *Journal of the Royal Geographical Society of London*, *20*, 1.

Lock, G. (2010). Representations of space and place in the humanities. In D. Bodenhamer, J. Corrigan, & T. M. Harris (eds.), *The spatial humanities: GIS and the future of humanities scholarship* (pp. 89–109).

Luchetta, S. (2017). Exploring the literary map: An analytical review of online literary mapping projects. *Geography Compass*, 11(1), e12303.

Mostern, R. & Johnson, I. (2008). From named place to naming event: creating gazetteers for history. *International Journal of Geographical Information Science*, 22(10), 1091–1108.

Papacostas, Ta. (2013). IBCC: Home. Available at: http://ibcc.dighum.kcl.ac.uk/ (retrieved 20 June 2018)

Ristow, W. W. (1967). *The emergence of maps in libraries.* Hamden, CT: Linnet Books.

Shirley, R. W. (2001). *Kitchener's survey of Cyprus 1878–1883 : the first full triangulated survey and mapping of the island.* Bank of Cyprus Cultural Foundation.

Simon, R., Barker, E., Isaksen, L., & de Soto Canamares, P. (2015). Linking early geospatial documents, one place at a time: annotation of geographic documents with Recogito. *E-Perimetron*, 10(2), 49–59.

Svensson, P. (2013). Humanities computing as digital humanities. *Defining digital humanities: A reader*, 159–186.

Weimer, K. H. (2011). The founding of ALA's Map and Geography Round Table: Looking back to see the future. MAGIRT Electronic Publication Series no. 11.

Zook, M. A. & Graham, M. (2007a). Mapping DigiPlace: Geocoded Internet data and the representation of place. *Environment and Planning B: Planning and Design*, 34(3), 466–482.

Zook, M. A. & Graham, M. (2007b). The creative reconstruction of the Internet: Google and the privatization of cyberspace and DigiPlace. *Geoforum*, 38(6), 1322–1343.

8 Motion in place

Introduction

Thus far this survey of the spatial humanities has explored the origins of the spatial turn (and by extension, the origins of the spatial humanities itself) in the fields of web history, textual scholarship, archaeology and information structures (gazetteers); and it has looked at spatial narratives, and the formal and informal data information structures that accommodate them. The non-neutral nature of place in the humanities (and the research concerned with it), and the potential of digital methods to act as crossing points between these otherwise disparate areas where they touch on place-making and place mediation, have likewise been highlighted. However one final area which requires survey before any general conclusions can be drawn is that of humans' *physical* engagement with place, and with the physical environment, through the agency of movement. For much of its history, one of the implicit purposes of the WWW has been to collapse physical space – essentially to dematerialize it. Cognitive perceptions of place and the development of communicative media are closely linked, and the spatial turn is a product of the shifting perspectives of content and scale that digital place has engendered. This broad understanding underpins the four frameworks of spatial narrative in Chapter 6 – they represent classes of the types of place that arise through communication between individuals and groups.

This chapter examines recent work on the importance of physical dynamism and movement in the spatial humanities. It looks at the way in which digital methods can be used to understand and explore physical dynamism in the present and how it is important to place-making, and how these methods can be projected to movement in place in the past. This revisits a tension that has surfaced and resurfaced throughout the preceding chapters – the basic truth that human interaction with space is analogue in character, yet freezing that interaction in a form that a computer can read is a form of digitization, and thus by definition is not analogue. Three such methods which have emerged from the spatial turn since the mid-2000s have proved particularly important. These are Qualitative GIS, which offer formal ways of rendering 'fuzzy' human data into quantifiable and mappable forms (e.g. Kwan & Ding, 2008);

motion capture technologies, which capture 3D traces of human movement in real time, which allow researchers to conduct post hoc analyses on how humans interact with different kinds of space at the micro-level; and finally the continuing rollout of GPS (and related GeoWeb technologies) into daily use in portable devices. With the trajectory of GeoWeb history described in Chapter 2, this is a developing phenomenon in the context of contemporary digital culture. The ubiquity of GPS in (developed western) societies means that for the first time since the spatial turn focused attention in the humanities on the digital study of place, humans are now, in fact, *living* in digital place in real time.

The physical environment itself is also a medium of communication, and human movement represents an *embodied* form of spatial narrative. Azaryahu and Foote (2008) explicitly differentiate between spatial narrative and narratives contained in the kinds of storytelling media which are normally the subject of narrative research, because the former is made in a specific location, and the other is not. While this perspective differs slightly from the argument offered here, the specific properties and philosophies of physical movement and how these relate to the disembodied WWW are certainly important to narrative-making. Most obviously, the kinds of developments in the WWW, the shift towards ubiquitous mobile computing (as described in Chapter 2), and the advent of haptic and motion capture technologies, have destabilized the 'disembodied' nature of the WWW as a means for linking only 'disembodied' information. It is no longer purely an 'information super-highway', with information delivered only to the desk top; rather it now sits alongside many of our physical interactions with the physical world around us.

Embodied and disembodied place: the tensions of cartography

Maps are representations of stasis, but – as described in Chapter 4 – they also *create* stasis. They provide a fixed, unyielding view of the physical environment, and of 'significant' features within it – whatever significance means. Once represented, and once read by a reader of a map at any one point in time, they become a freeze-frame representation of the landscape or cityscape at that point. The perception of stasis is reinforced and entrenched with every (re)printing or copying of the map, and with every (re)reading of it (the ekphrastically embodied nature of reading highlighted in Chapter 4, has some parallels with this) but, picking up on an established theme of this book, these points of reading translate place in which movement happens from discursive chorography into measurable geography. The inimicalness of maps for representing the subjective dynamic and temporal fluidity – and the fact that these are so important in the past and the present – masks this process of translation. Rediscovering the importance of movement in the human record, in turn, provides much of the motivation for the spatial humanities.

On the other hand, the deep mapping/spatial narrative approach acknowledges that space is not static; and that while the statement that everything happens somewhere is over-used to the point of being anachronistic (Warf & Arias, 2008), space and location drive human action and events, as post-processual approaches to archaeology are acknowledged explicitly. The introduction to a key recent volume on this describes mobility as 'an ensemble of freedom, opportunity, adventure, progress, and yet … was also a form of restriction'. (Leary, 2014: 16). The creation of spatial narratives, informed by post-processual theory, can only stem from embodiment: an understanding of what it is to move through a landscape, a building, a site-specific performative space. This is a perspective that draws heavily on performance theory; and there have been interdisciplinary attempts to bridge the dichotomy (Pearson & Shanks, 2001), an otherwise empty space which is encultured by human presence.

Like representational maps, the standards-mediated depictions of place through the affordances of the digital realm also perpetuate a static, frozen-in-time and disembodied view of the past. We can create a detailed point-based gazetteer of an ancient society as detailed in Chapter 7, and use web-readable URIs to give each place a unique identity; but this takes us little further in understanding how mobility, perambulation and embodiment contributed to those societies. We thus fail to address the kind of phenomenological challenges articulated by Tilley and others, and the notion of 'dwelling' within a landscape, as opposed to mere population or presence. This, it might be argued, is also a product of the medium of the web, which by definition separates the person (the 'user') from the information, never mind from the place in question.

The lack of a critical approach to human movement through space, combined with the separative properties of the WWW, creates the same kind of critical vacuum which is filled – to the detriment of balanced critical enquiry, as described in Chapter 5 – by the neogeographic GeoWeb; or by institutions and organizations which already have cultural power. Take the case of Western study of Middle Eastern cultures. In 1978, the cultural theorist Edward W. Said critiqued the objectification of 'the orient' by western scholarship by labelling it 'orientalism', and writing in his classic work of the same name:

> The Orient and Islam have a kind of extrareal, phenomenologically reduced status that puts them out of reach of everyone except the Western expert. From the beginning of Western speculation about the Orient, the one thing the Orient could not do was to represent itself. Evidence of the Orient was credible only after it had passed through and been made firm by the refining fire of the Orientalist's work. (Said 2003: 283)

The idea of 'phenomenological reduction' is critical here. It refers to the observation that representation of cultural artefacts and heritage was only possible through the processes of abstraction and critical quantification of (in this case western) interpretative frameworks, frameworks that take precedence where there is *no* first-hand experience or memory, *no* physical engagements with the objects or their cultures, and *no* set of 'extrareal' critical frameworks.

This implicitly denies the corporeality of physical engagement, or the ability to exercise such engagement, at least in an uncurated manner. The WWW, arguably, for the reasons set out in previous chapters, accentuates this 'reduction', and leaves it open to the uncurated and potentially prejudiced processes of knowledge formation to which the (open) GeoWeb has become susceptible.

However, modern theorists, artists and practitioners who construct the praxes of movement, and what we might call the curation of movement through the physical environment *have* engaged with this in various ways, and it is through such engagements that the spatial humanities can establish critical and – perhaps – deep map-like frameworks to the mapping and understanding of encultured movement. Azaryahu and Foote, for example, draw a distinction between spatial narrative and the kinds of narrative conveyed through other media, such as film, literature etc. This requires that historical spaces of particular importance (which is to say they are given importance by historical events which occurred in, and shaped them) are curated to give prominence to those events, which can occur at particular scales. This process is obviously subjective – they highlight the difference, for example, between describing the site of a 'battle' versus the site of a 'massacre'.

Such a post-processual view is critical for the spatial humanities. The human landscape is a complex mixture of elements which have been designed and constructed in a tangible and top-down manner, and elements which result from intangible and unplanned engagement. The former might be roads, motorways, canals, and buildings – and the 'Archaeological Entities' of the Cyprus gazetteer described in Chapter 7. The latter are traces which are made without being planned. A 'line of desire' for example, which deviates from an 'official' or planned pathway, can be created when footfall repeatedly deviates from the latter, and along a pathway which is easier or more convenient for most users of the path to follow. In many ways these desire-lines occupy a moment of fleeting, transient (and chorographic) significance to the individual wayfarer, but most times they will not be picked up by any official planning processes. This is especially so at the smallest of scales, e.g. Figure 8.1.

The tensions between static and fluid marks on the landscape, between planned and unplanned traces (and from there it is a short etymological jump to 'mapped versus unmapped'), and how different media capture these traces, has proved a rich seam for visual artists. In 1967, the photographer Richard Long took a break from a hitchhiking journey between Bristol and St Martins and, in a field in Wiltshire, walked up and down until the grass was flattened by the repeated passage of his feet. When it caught the sun, a visible line was created, and Long photographed this. The result, *A Line Made By Walking* (now at the Tate) is an exploration of the engagement between ephemeral and disembodied traces of human passage through landscape, and the potential for the media of human discourse to extend – perhaps indefinitely – the temporal span of an otherwise time-limited moment of phenomenological engagement

Figure 8.1 A 'line of desire' deviating, at micro scale, from a designed and planned pathway.
Photo: the author

(Figure 8.2). The stasis and fixity of the photograph contrasts with the ephemerality of the line it documents. The image illustrates the tension between the fluidity of movement through the landscape, the transience of the traces it leaves, and the fixity of the photographic medium.

Such tension makes up our spatial reception of the world; and it is exponentially exacerbated by mass communication media (especially visual communication media). For Susan Sontag, photography represents the images in Plato's Cave (see Chapter 6), and the mass of frozen imagery on the wall of the Cave contains billions of instances of the ephemeral-static conversion exemplified by Long's line. They form a vast part of our collective visual consciousness.

A more recent example of this is 'GPS drawing'. In this, the artist uses GPS to capture his or her movement through the environment as a trace at 1:1 scale. This presents that movement set as a single entity, with an artistic significance that is only apparent when it is presented as a whole at the end. One of the pioneers of GPS drawing, the San Francisco-born artist Jeremy Wood, has particularly explored the potential of using his personal movement, emplaced and represented by GPS traces, in the production of artworks which challenge both scale and temporality (Lauriault, 2009). His *Traverse Me*, for example, is a 1:1 map of the campus of the University of Warwick, entirely observed and traced by him walking the lines of the map and recording them

Figure 8.2 A Line Made By Walking by Richard Long.

using GPS. This challenges and subverts all the temporal and scalar assumptions of representational cartography. *Traverse Me* is, essentially, an entirely chorographic map, drawn from personal perambulation, yet it transfers seamlessly into the digital world, with no loss through recalling or reinterpretation.

Movement, the personal experience of movement, and physical engagement with the environment around us, are all closely connected; and, as has been seen in previous chapters, the recording and sharing of spatial information does not necessarily transport well across time. This includes the documentation and recording of movement. The nexus between personal experience and observation of a landscape through movement, and an understanding of that landscape's human-cultural significance is a core philosophical concern of phenomenology, whose place in the spatial humanities has been discussed previously (see Chapter 3 especially). Most of these debates concern the understanding of the past by physically experiencing the environment of the present. Spatial narratives born of such experience gain legitimacy, or not, from the way in which they contribute to particular kinds of interpretation, and how these relate to what the viewer observes and records. The orientation of a long barrow, for example, or the restriction and allowances of sight when ascending a hillfort have an interpretative significance which combines the viewers' sensory experience with information from other sources. In one sense therefore, phenomenological approaches to landscape face a challenge similar to that of researchers seeking to understand a historical literary allusion in a text: the 'readers' are both projecting their spatial understanding back in time, and seeking a critical, evidence-based understanding of what that space meant to those who occupied it then. The former 'reader', however, is connecting with the contextual information of the place in question, as described and mediated in the text, the latter is connecting with the place itself.

The event-based 'beginning, middle and end' model of narrative clearly does not work for some non-Western contexts, which might be far removed from the structuralist narrative tradition. The cultural importance of some landscapes simply cannot be elaborated through discursive and dis-locative media (such as text or image). One example, which highlights particularly the deep division between landscapes derived in text and landscapes experienced in person, concerns Australian Aboriginal rock art. The ethnographies of the Aboriginal societies conceive of mythic creation traditions rooted in the movement of supernatural beings through and around the landscape. In his classic 1987 travelogue exploring sacred Australian societies *Songlines*, Bruce Chatwin describes the way in which the lines connect these spiritual events: 'the melodic contour of the song describes the nature of the land over which the song passes' (Chatwin, 1987: 108).Topographic features such as mountains, streams, woods, sources of water and food – the fundamental elements of landscape that predicate survival in a hunter/gatherer society – are formed by them as they engage with that landscape by moving, stopping, sitting down, hunting etc. Personal, individual experience of these features in the present is irreducibly connected with these processes. Often, sites of significance in this narrative are marked by rock art, whose (apparently) stationary and immovable nature (a form of 'monumentality' according to Western traditions) belies the complex processes of movement and observation behind its creation. As Ursula K. Frederik has noted, '[B]ecause it facilitates the ongoing reproduction of relations between humans and the places they share, rock art might … be acknowledged as a locus of moving ideas in its own right' (Frederick, 2014: 65) She goes on to point out that rendering rock art on static 'grid maps' risks imposing and perpetuating 'metaphors of post-colonial dominance': 'telling' place according to Western narrative structure, and its attendant Cartesian/positivist understandings of geography, location and place. This removes the original meanings of the art from its context, thus changing its meaning.

Wallis (2009) further develops this theme in his discussion of rock art from the point of view of 'animic ontologies', which rejects an explicit distinction between 'human and non-human persons', between (human) culture, including material culture embedded in the landscape, and nature. In this schema, non-human features of the landscape, which may thus be anthropogenic (and thus including the four aspects described above) can be imbued with 'human-like' qualities: 'rather than imposing meaning in the form of nonhumans onto the landscape in a psychological projection or attributed anthropomorphism, this represents a complex web of relations' (Wallis, 2009: 55). It is true that such an analysis risks a circular logic, in that the nature of the 'nonhumanness' must inevitably be a product of human cognition; however Wallis also (rightly) rejects the notion of a grid-map approach to 'nonhuman' rock art, stating 'Lived-in-landscape is not a singular entity, contained by the gridlines of a map or reproduced in pixelated three dimensions by sterile virtual imaging' (ibid.). It also highlights a risk inherent in using any narrative strategy to explain the processes of the cultural elaboration of landscapes, including spatial narrative strategy, namely that of imposing a narrator's viewpoint, as the viewpoint of a cartographer is inevitably a subjective and imposed viewpoint of the terrain being mapped. While the web, and its standards, might restrict the formation of spatial narrative, its affordances for allowing anyone – or at least anyone with an internet connection, which is of course not everyone – to add to the landscape of geodata inevitably raises the question of who the authors of a collective deep map can be, and of what the process of authorship means. The issue of collective authorship of deep maps is returned to in the concluding chapter.

Shaping the landscape

Motional narratives are thus created in the past and recreated in the present, mediated by specific cultural circumstances, and reflecting the truth that tactile and bodily movement through the landscape happens in both the past and the present. Not only does this shape human experience through the cultural lens of the observer, as above, these processes of physical interaction also shape the landscape itself. As with the conceptual landscapes of neogeography, there are both designed, 'top-down' ways in which this happens, where there are clear designs and motivations – the building of a new motorway for example – but there are also bottom-up, collective means through which movement shapes the landscape in an undesigned way. Any discussion of the physicality of movement must begin with the way in which the processes and significances of collective human movement leave their mark on the landscape itself. Urbanized, or even semi-urbanized human culture lends itself to creating cross-country courses in order to facilitate communication, whether that be the physical movement of people, or the movement of goods, or, more latterly, network of communication. The nature writer Hugh Warwick calls these 'linescapes', lines in the landscape which not only connect points of significance, such as towns, but at the same time fragment the landscape (Warwick, 2017).

The impact and visibility of linescapes on the landscape is, most often, a function of the volume of movement which they facilitate. This is reflected in Jeremy Wood's work, as discussed above, where a thicker or darker GPS trace may represent a line of walking which is more heavily used than a lighter line (Lauriault, 2009). However, un-designed traces reflecting usage also emerge over the real-time of centuries.

One of the oldest, as well as one of the most literal examples of this, most visible in temperate regions of chalky or clayey soils, is the so-called 'green lane' or hollow-way. These are deep trackways scored in the soil by the passage of feet, hooves and wheels over centuries, or even millennia. 'Green lanes' are deeply evocative places. Their courses are determined by the landscape, because the impetus upon travellers to minimize the physical and economic costs of travel, the action of gravity, and erosion by wind and water will all play their part in defining the route they choose to take. The cumulative impact of decades or centuries of their footfalls etches the line further, (literally) entrenching it. Once established, they act as agents which shape the landscape, their depth testifying to usage through movement. In his book *Holloway*, Robert Macfarlane describes them as '[l]andmarks that speak of habit rather than suddenness. Like creases in the hand, or the wear on the stone sill of a doorstop or stair, they are the result of repeated human actions' (Macfarlane, Donwood, & Richards, 2012: 3).

Yet the cartographic and philosophical limitations of documenting movement described above militate against the systematic study of the history of green lanes. One of the key problems in the study of old trackways as historical phenomena is fact that the historical presence of movement is the principle, if not the only, source of evidence. One of the first systematic surveys of green lanes, R. Hippsley-Cox's *The Green Roads of England* (1914) noted that '[m]uch is, and must be, guesswork, since all the evidence that remains to guide us are the trackways and the earthworks' (Hippsley-Cox, 1914: viii). Many green lanes are bordered by hedgerows; and one rule of thumb, is 'Hooper's Hedge Hypothesis', which states that the number of species of trees and woody shrubs within any 30-metre length will, approximately, equate to the number of centuries the hedgerow has been in existence (Hooper, 1970). The absence of any more grounded historical evidence for the qualitative, or experiential, use of green lanes has, historically, left the door open for speculative and/or unempirical interpretations of landscape history. Hippsley-Cox himself, for example, posited, on the basis of the morphology and geography of the green roads around key sites such as Avebury, that the 'civilization' of the pre-Bronze Age period of southern Britain was more advanced and complex than that which followed, and supported a form of centralized government – a theory of prehistory that no modern interpretations based on excavation done since would support.

Case study: corpse roads

While it is extremely difficult to extract historical or archaeological ground truths from rural trackways, and the significance of movement along them, in some contexts it is possible to discuss certain aspects of this significance; where there are other kinds of supporting evidence, from history perhaps, or folklore, or archives. In this, we see once again the transmedial nature of spatial narrative discussed in the previous chapter.

One example of this is the role of trackways and routeways in the context of death. This is particularly interesting as an example, because most religions and cultures have processional activities associated with the removal of a body between the place of death, to a place of ritual, to the site of burial or creation (if different). In this way the network(s) of route(s) thus taken become imbued with contemporary cultural meaning as a result of the passage both of the corpse and of the funeral participants. In certain contexts, and with the availability of certain kinds of evidence, digital methods can been used to explore and understand these processional events. Such an understanding is necessarily dependent on a full picture of the context of the phenomenological and experiential engagement of individual participants with the environment, with whatever empirical evidence is available, and with the topography of the route itself. These factors such as gradient and terrain, what one can see from particular points, all aggregate together into a holistic understanding of the route both as a *space* and a *connector* (see Vitale, 2017). For example, Favro and Johanson's analyses of funeral processions through the Imperial-period Roman Forum using 3D environmental reconstruction established that the Severan construction programme reconfigured funerary routes using propagandistic inscriptions and sculpture. These redirected both the routes taken by the processions and the gaze of the participants while they traversed them (Favro & Johanson, 2010: 32). Both of these are, of course, functions of trajectory – they only make sense in the context of phenomenologically engaged individuals and groups traversing the space, which is conditioned by the architectural environment.

However, such culturally and/or religiously charged contexts of traversal can also imbue the routeway *itself* with cultural or religious significance, manifesting itself in the perspectives of religion, folklore and superstitious belief. The latter particularly are evidentially and interpretatively problematic, as they often lack formal written records or artefact evidence. One good example which highlights both the challenges and the opportunities of a transmedial approach to such phenomena are the beliefs associated with so-called 'corpse paths' in the Christian West. During the later medieval and early modern period, before the extension of the metalled road network beyond trunk routes connecting major population centres and the Enclosure Acts (Hindle, 1982), an inevitable result of a death in a remote village would be that the corpse, whether placed in a coffin or wrapped in a shroud, would have to be carried to the place of burial. Corpse paths were thus routeways taken, often over rural landscapes, by funeral parties taking the corpses of the deceased from their place of death to the place of burial. As the Church consolidated its social and economic power by insisting that deceased communicants be buried in consecrated ground, and the consequent development of the churchyard after the tenth century (Zadora-Rio, 2003), this led to the development of corpse paths, or corpse roads, which were supposed to provide a symbolic connection between the place of death and the burial. The belief was known to Shakespeare who, in *A Midsummer Night's Dream*, has Puck soliloquize: 'Now it is the time of night / When the graves all open wide / Every one lets forth its sprite / In the Church-Way paths to glide' (Act V, Scene 1).

There are some direct references to corpse paths in antiquarian literature dealing with historic folklore and oral traditions. In *Odd Yarns of English Lakeland*, W. Palmer writes that

> Some of our mountain hamlets are far from the parish church, which has given rise to the "corpse road," which goes straight a lance to the village centre … But the official who dared to meddle with the corpse road, even though it might not be used once in twenty years, was in for dire trouble. (Palmer, 1914: 37).

Another aspect of corpse path folklore with, arguably, a more robust historical grounding, is the belief that if a party bearing a coffin crossed private land it created a public right of way – an (erroneous) belief which persisted until relatively recent times. This is a belief addressed by William Weeks, in a short paper in *Folklore* in 1928. Weeks quotes the 1562 *Book of Homilies*:

> that where their ancestors left of their land a broad and sufficient bier balk to carry the corpse to the Christian sepulchre, how men pinch at such bier balks, which by long use and custom ought to be inviolably kept for that purpose: and now they either quite ear [plough] them up, and turn the dead body to be borne farther about in the high-streets; or else, if they have any such meer, it is too strait for two to walk on.
> (Weeks, 1928: 394–5).

This thesis articulates a belief that a very particular type of movement through the landscape, the bearing of a corpse, gave rise to the practice of leaving a strip of land unploughed at the edge of the field. One example of such a path is explicitly referred to in this 1928 paper, linking Littleworth in present-day Oxfordshire with the burial site of the abbey at Faringdon, approximately two miles to the west. Examination of the present-day footpath corroborates the inference that the path connecting the two settlements did indeed have some corpse-road-like associations, as its contemporary name (indicated on signposts) is 'Church Walk'. The present-day topography also gives some clues. For approximately half a mile from Faringdon, there are hedgerows which appear to be of some antiquity (Figure 8.3) either side of the path, which runs between ploughed fields. This would appear to offer circumstantial support for the idea that a strip of land had

Figure 8.3 Hedgerows on 'Church Walk', near Faringdon, Oxfordshire, UK. Photo: the author

been left in one or other of the fields for the purpose of passage – whether for the passage of a corpse or not is unclear. However, the fact that these are only in evidence near the site of the burial ground suggests that – *if* they do represent strips of land left deliberately for the purpose of allowing bier parties to pass – then such practices were only in evidence near the place of interment.

Therefore, corpse paths of the kind described by Weeks, or the 'church-way paths' of Shakespeare only come into being through the agency of movement close to the burial ground itself. Paths over countryside served a variety of purposes, and had a variety of significances, and this brief case study shows that by drawing on a range of archival and observational sources, one can at least infer how human passage, and the conjoined movement of life and death through a landscape, can enculture that landscape with meaning.

It must be stressed very clearly that this corpse roads case study is as important for what it is not, as for what it is. It does not, for example, represent any kind of an exercise in scientific landscape archaeology in the sense discussed in Chapters 3 or 6, as the empirical evidence that would be needed to underpin such an exercise is not there. For the reasons set out in critiques of phenomenology, and the responses to Tilley's view of physical and embodied engagement as a method for understanding the past, the cognitive jump involved could not be justified according to the kind of empirical standards required for an interpretation of the past that would be justifiable or replicable, according to the standards of academic knowledge discussed in Chapter 5. What the case study does provide however is a creative means of thinking about place in the past, and draws on other disciplines and epistemologies through the spatial significance of movement, and the way in which movement adds cultural significance to the landscape – and vice versa. As Brück puts it, 'perhaps one of the most productive strands of phenomenological writing within archaeology has been the deconstruction of the dualistic thinking that is a product of post-Enlightenment rationalism' (Brück, 2005: 65).

Case study: Experimental archaeology and motion

Brück's view suggests that movement, and phenomenological engagement, does not sit easily with the hyper-materialist approach to cultural heritage, which tends to relegate the human element to a position below material interpretation, placing firm emphasis on the importance of tangible evidence. This opens up important questions about the role of contemporary narrative structures in the interpretation and understanding of past movement. It is a mode of thinking that reached an apotheosis in the seventeenth and eighteenth centuries, an era when (in the European and Anglophone spheres) the encyclopaedic museum was developing – the British Museum, for example, was established in 1753, and Washington's Smithsonian Institution in 1846. Such institutions projected an idea of the universality of human culture, as represented in its material objects (Westerhoff, 2001). The only relationship such collections had to embodied, or phenomenologically engaged movement, was where visitors moved through the collections, observing, receiving and understanding them according to the narratives and structures of the contemporary curated environment. In some cases, such as with the Great Exhibition of 1851, these were overtly colonial in character (Coombes, 1986).

This approach to materiality lies behind many assumptions about the nature of cultural heritage. Ideas about scientific typology were inherently bound up with broader intellectual movements. While, for example, Darwinism detached the fate of man from the fiat of God (see discussion in Chapter 3), and ushered in an age of understanding of 'deep time', it also provided a framework for the understanding of human antiquity that was conditioned by progression from the simple to the complex – what Gamble and Moutsiou (2011) call 'a well-stratified modern mind with a primeval basement' (see also Dunn, Earl, Foka, & Wootton [forthcoming].for further discussion). This was based on systematically constructed sequences of culture, founded on the principles of evolutionary theory. In such a view, one class, or type, of material culture – such as a category of pottery or arrowhead – mutated inexorably into another, the later one more sophisticated that the earlier. No kind of embodied or physical engagement with the contemporary spatial context of culture itself was necessarily required in order to understand this process; and indeed any attempt to introduce discussion of processes which are *not* material, measurable and rational was seen as problematic for the scientific assertion of factual narratives about the past.

This reflects long-standing debates in experimental archaeology, a branch of archaeology which contains important context for these questions, and for human movement mediated through digital environments. Experimental archaeology involves understanding the processes of creating artefacts from the past by replicating both the process and the artefact in the present, thus helping to understand the dynamics that lay (and indeed lie) behind its manufacture. Reflecting the debates in phenomenology described above, and in Chapter 3 especially, some pioneers of experimental archaeology were very careful to separate the material from the immaterial in this, arguing strongly that there is no intellectual merit at all in attempting to situate human activity in a reconstructed material environment. Peter Reynolds, the founder and first director of the Butser Ancient Farm site in Hampshire, England, where a number of Iron Age roundhouse structures have been physically inferred and constructed from archaeological evidence, wrote:

> In real terms it is only sensible to examine structures physically and as far as possible to dehumanise the examination process. Reenactment is best left as a dramatic indulgence to the imagination, which can be recognised as singularly valueless and instantly forgettable … History, and by implication prehistory, is swiftly becoming a tabloid newspaper subeditor's view of the past,
>
> (Reynolds 1998)

Reynolds's use (and criticism) of the word 'reenactment' is significant here. In (spatial) narrative terms, this implies a conscious effort to re-create an intangible event which happened at a particular moment in historical time. A

Civil War battle may be reenacted, for example, but the human processes involved cannot be empirically reconstructed as they happened. Rather they are re-interpreted through the contemporary person's embodied response to the landscape, its stimuli, and (in the discourse of Wallis and Fredrick above), any human or non-human actors present in it. There can be no empirical or rationalist way of connecting them.

Museums and other managed cultural spaces approach the values of movement in the present rather differently. Unlike Reynolds's approach to experimentalism, whose only aim is systematically to construct empirically viable material structures in the present day, and very clearly, avoid imbuing the intangible processes involved with interpretative significance by 'curating' it in the physical context of Butser, museums explicate the idea of curated movement, sometimes intentionally, sometimes implicitly. Institutions associated with the preservation and production of collective or social memory, particularly when they have a 'national' role to play, adopt very specific approaches to the location and operationalization of movement-based spatial events, and the spaces within them may often be 'curated' to the same degree as the objects in the cases, and they enable The British Museum in London, for example, is a location which embodies a narrative about the importance of the material cultural collections it contains. Its soaring neoclassical portico, the ascent to the front entrance up a flight of 12 steps, the sense of light created by the glass roof of the Great Court provides a narrative of authority. This is in turn conferred on the curatorial narratives of the collections – (at least from the official perspective) one of ownership for the nation, and providing access for all.

The museum's spatial footprint can be represented as a grid map, and the pathways of its visitors plotted around the galleries and public spaces using the same geodectic standard. We can note that the British Museum can only be reached (recalling Tilley's emphasis on the importance of movement through the landscape, in this case an urban one, being as important as arrival) by travelling through the landscape of Bloomsbury, with its connotations of wealth and establishment. But this monolithic narrative provides the context for many 'smaller' narratives, which similarly make sense only in the spatial context of movement through the museum's space(s): those of the visitors and staff, their reactions and interactions, how it shapes their perspective of the landscape and the wider world; and in the digital traces they leave online as a result of those responses. Spatial narrative may also therefore be embodied by people, in the formation of 'maps of inner terrains' as Macfarlane puts it, arising from experience of, and interaction with, the spaces and cultures around the individual. It provides context to the empirical materialism of the physical collection, rather than seeking to replace it.

It is also the case that the type of spatial narrative facilitated at (and by) movement through historical sites changes over time. This forms a curatorial basis for the presentation of history and heritage at those sites. For example, as the site of a Roman fort complex at Arbeia, South Shields, fell into disrepair, walls and other dividing structures which previously restricted and contained movement became mere guidelines. This gives rise to narratives at the site which are personal to the visitor; they can also be generic in nature. The site consists of a mixture of foundation walls, curated as they have been preserved. Other parts, such as the West Gatehouse have been reconstructed to two or more storeys from a variety of sources (Figure 8.4). The sensory experience of moving in through the site thus recalls the idea of the deep map, but the walls, now only foundations, direct the visitor/experiencer through the space. The walkthrough nature of the reconstruction and curation, and the way these guide the walkthrough, still constitutes a full sensory environment. However, like the corpse road example, this is not 'scholarly' in the sense that it does not provide an evidence-based account of what it would have been like for a first- or second-century Roman, or auxiliary soldier to experience the environment.

The notion of reconstruction at Arbeia is therefore multifaceted. It is partly concerned with the physical reconstruction of buildings, and partly with offering interpretations of the past by conditioning visitors' sensory engagement with the environment in the present. This invites us to situate the 'creative thinking' of Brück (2005) with the empiricisms of Reynolds (1998), and to consider how the third class of spatial technologies referred to at the start of this chapter, motion capture, might help bridge the gap between the two.

Motion capture has a significant role to play in the 'archaeology of mobility'. It allows human movement to be recorded in 3D and in real time, has a range

Figure 8.4 The site of Arbeia, South Shields.
Photo: the author

of applications in dance and performance (van der Linden, Schoonderwaldt, Bird, & Johnson, 2011; Chan, Leung, Tang, & Komura, 2011), video games (Menache, 2000) and health (Park, Park, Kim, & Adeli, 2015). In the early 2000s, it was bought to widespread public prominence through its role in major CGI movie projects, such as the *The Lord of the* Rings trilogy (Scott, 2003). A project co-led by the author in 2010–11, and funded by the AHRC's DEDEFI programme, sought to explore its applications in the understanding of cultural spaces. This used motion capture to analyse responses of different people in the present day to re-constructed archaeological environments (Woolford & Dunn, 2013; Woolford & Dunn, 2014). A number of historic environments were explored, but those most relevant to this discussion were the interiors of Iron Age roundhouses reconstructed through the methods of experimental archaeology at Butser Farm (Reynolds, 1979). While there is very little written testimony for daily life for the inhabitants of Iron Age Britain, archaeological evidence indicates that these environments would have constituted day-to-day living quarters for people at various levels of society, including members of non-elite communities. This means that they are out of the archaeological and evidential range of the kinds of groups, examined, for example, by Favro and Johanson in their work on Roman funerary processions.

The environments clearly present a number of navigational challenges for one not used to them – the inner ring of posts compromises visibility; the floor is uneven, and there is limited space for movement near the inner edge where the (thatch) roof descends towards the wall (Figure 8.5). The aim of the

Figure 8.5 Interior of reconstructed roundhouse at Butser, England.
Photo: The author

experiment was to measure the navigational reactions of persons familiar with the place against those who were not, thus elucidating how reactions change with the acquisition of spatial memory. As it was conducted within the context of an experimental archaeological reconstruction which had met the standards of material exactitude set by Reynolds (see above), it was considered that assessing contemporary human responses in this way was simply an extension of that experiment. Whilst there can be no *linear* comparison with the ways an Iron Age person and a twentieth-century person navigated the space, the visualization nonetheless helps us to 'think creatively', in Brück's words, about the principles of phenomenological engagement. It extends the materiality of experimental archaeology to *direct* experimental engagement, with the material.

Two sets of participants, student volunteers and experimental archaeologists who had worked with the spaces at Butser for a considerable period were equipped with gyroscopic motion capture suits, and asked to perform a set of domestic tasks with the environment. In one of these, operating a quern stone, there is a marked difference between the patterns of movement of experienced and inexperienced users (Figure 8.6). More research will be needed to quantify the exact structures of these motion traces, and determine if there are any general principles that can be applied to their evolution over time as the 'phenomenological engagement' between the individual and the site increases with experience. Another obvious problem with the experiment is that two groups of different people were used in the study – a better model, with more time and resources available, would be to capture the movements of the same people over a period of years, and study the evolution of the traces in a longitudinal manner.

However, the data we have does allow us to visually 'fit' intangible assets, such as knowledge, practice and expertise into the markedly tangible context

Figure 8.6 Digital 3D reconstruction of an inexperienced agent querning (above), versus an experienced one (next page)

Figure 8.6 (continued)

of a physically reconstructed ancient environment. It is a very rudimentary example of the digital medium providing a bridge between tangible and intangible ways of thinking.

Conclusion

This hints at some general conclusions that can be drawn about the nature of digital remediation. The 'remediation' of cultural heritage, be it physical objects or intangible movement, involves the creation of digital objects which represent the non-digital world. In the creation of any digital object, certain processes must be gone through. Most important is the need for transparent and consistent documentation of the process, so that future audiences can understand how it was created. The process of creating a 3D digital visualization of an artefact can be tracked, documented and thus made replicable, and there are established formal structures which enable this. Most notably, the London Charter is a standard which recommends good practice in the building of 3D models, spread across six 'principles' of implementation, aims and methods, sources, documentation, sustainability and access (Beacham, Denard, & Niccolucci, 2006). Most importantly perhaps, the materiality of the original – if it is available for comparison – provides an obvious and empirical standard for comparison. The remediation of spaces, buildings and monuments entail similar issues (Johanson, 2009). Architectural reconstructions can be more complicated to remediate, precisely because they are facilitators of movement (in the way the remains at Arbeia are, as discussed above). More recently Vitale (2017) has sought to construct a set of principles for documentating the construction the remediation of archaeological monuments and spaces in a manner that is amenable to exposure and manipulation in Linked Open Data structures; and which takes account of the fact that the occupants of buildings now not extant would have moved from one room to another. For this reason, Vitale's approach places more emphasis on the ontological structure of the data behind the reconstruction than its visual properties, meaning that relationships (including the kinds of connective relationships that facilitate and enable movement) can be expressed as links.

Digital materials are a product of the processes of remediating both movement and objects; however objects and movement present different problems in terms of the historical veracity of those digital materials, and the evidence on which they are based. This evidence takes various forms. While remediating movement and remediating objects therefore both result in the construction of digital materials, they both present different problems in terms of how such structural processes of remediation function alongside the process of object creation. remediated archaeological objects. It is useful to express this commonality in terms of the certainty of each type of evidence. First, there can be empirical statements. It is easy, for example, to make an empirical statement that a piece of Roman Samian Ware pottery is red in colour and is 5 centimetres in length. Another empirical statement might concern the reconstruction of an Iron Age roundhouse, and state that 'the inner ring of this structure comprised of twelve upright posts, because we can discern twelve post holes in ring formation'. This is the kind of grounded, ultra-rationalist interpretation that Reynolds argued for at Butser (see above). Another type of conclusion might be based on *interpretative* processes. It might be inferred, for example, that our notional piece of Samian Ware possessed a handle at some point in its history, due to comparison with other vessels of the same type and date, and inferences about its function. In the same way, we might conclude that because ceramic sherds were found in a post hole in a roundhouse, cooking and/or eating took place near to this point. Finally, there are *conjectural* conclusions that might be included in the digital object. This could include a conjecture that a particular kind of meat was cooked in a particular way at this location, based not on interpretation or empirical evidence immediately to hand, but on the general context of the environment, and on what is known more broadly about domestic practice.

References

Azaryahu, M. & Foote, K. E. (2008). Historical space as narrative medium: on the configuration of spatial narratives of time at historical sites. *GeoJournal*, 73(3), 179–194.

Beacham, R., Denard, H., & Niccolucci, F. (2006). An Introduction to the London Charter. In The E-volution of ICTechnology in Cultural Heritage, Papers from the Joint Event CIPA/VAST/EG/EuroMed Event, 2006.

Brück, J. (2005). Experiencing the past? The development of a phenomenological archaeology in British prehistory. *Archaeological Dialogues*, 12(1), 45–72.

Chan, J. C. P., Leung, H., Tang, J. K. T., & Komura, T. (2011). A virtual reality dance training system using motion capture technology. *IEEE Transactions on Learning Technologies*, 4(2), 187–195.

Chatwin, B. (1987). *The Songlines.* Open Road Media.

Coombes, A. E. (1986). Museums and the formation of national and cultural identities. *Oxford Art Journal,* 11(2), 57–68.

Dunn, S., Earl, G., Foka, A., & Wootton, W. (n.d.). Spatial narratives in museums and online: The birth of the digital object itinerary [forthcoming].

Favro, D. & Johanson, C. (2010). Death in motion: Funeral processions in the Roman Forum. *Journal of the Society of Architectural Historians*, 69(1), 12–37.

Frederick, U. K. (2014). Suspended animations: Mobilities in rock art research. In *Past mobilities: Archaeological approaches to movement and mobility* (pp. 63–78). Farnham: Ashgate.

Gamble, C. & Moutsiou, T. (2011). The time revolution of 1859 and the stratification of the primeval mind. *Notes and Records of the Royal Society*, 65(1), 43–63.

Hindle, B. P. (1982). Roads and tracks. In L. Cantor (ed.), *The English medieval landscape* (pp. 193–217). London: Croom Helm.

Hippsley-Cox, R. (1914). *The green roads of England*. Glastonbury, England: Methuen.

Hooper, M. (1970). Dating hedges. *Area*. The Royal Geographical Society (with the Institute of British Geographers).

Johanson, C. (2009). Visual resources : An international visualizing history: Modeling in the Eternal City. *Direct*(June 2012), 37–41.

Kwan, M.-P. & Ding, G. (2008). Geo-narrative: Extending geographic information systems for narrative analysis in qualitative and mixed-method research. *The Professional Geographer*, 60(4), 443–465.

Lauriault, T. P. (2009). GPS tracings – personal cartographies. *Art & Cartography Special Issue*, 46(4), 360–365.

Leary, J. (2014). *Past mobilities: Archaeological approaches to movement and mobility*. Ashgate Publishing.

Macfarlane, R., Donwood, S., & Richards, D. (2012). *Holloway*. London: Faber & Faber.

Menache, A. (2000). *Understanding motion capture for computer animation and video games*. Morgan Kaufmann.

Palmer, W. T. (1914). *Odd yarns of English Lakeland*. London: Skeffington & Son.

Park, S. W., Park, H. S., Kim, J. H., & Adeli, H. (2015). 3D displacement measurement model for health monitoring of structures using a motion capture system. *Measurement*, 59, 352–362.

Pearson, M. & Shanks, M. (2001). *Theater/archaeology*. London and New York: Routledge.

Reynolds, P. J. (1998). *Butser Ancient Farm: Occasional Papers*volume IV. Waterlooville: Butser Ancient Farm.

Said, E. (2003). *Orientalism*. London: Penguin Books.

Scott, R. (2003). Sparking life. *ACM SIGGRAPH Computer Graphics*, 37(4), 17–21.

van der Linden, J., Schoonderwaldt, E., Bird, J., & Johnson, R. (2011). MusicJacket—combining motion capture and vibrotactile feedback to teach violin bowing. *IEEE Transactions on Instrumentation and Measurement*, 60(1), 104–113.

Vitale, V. (2017). *Rethinking 3D digital visualisation: from static visual aid to multivocal environment to study and communicate ancient cultural heritage*. Unpublished PhD thesis, King's College London.

Wallis, R. J. (2009). Re-enchanting rock art landscapes: Animic ontologies, nonhuman agency and rhizomic personhood. *Time and Mind*, 2(1), 47–69.

Warf, B. & Arias, S. (2008). *The spatial turn: Interdisciplinary perspectives*. Routledge.

Warwick, H. (2017). *Linescapes: Remapping and reconnecting Britain's fragmented wildlife*. London: Square Peg.

Weeks, W. S. (1928). Public right of way believed to be created by the passage of a corpse. *Folklore*, 39(4) (Dec. 31) pp. 393–398.

Westerhoff, J. C. (2001). A world of signs: Baroque pansemioticism, the polyhistor and the early modern Wunderkammer. *Journal of the History of Ideas*, 62(4), 633–650.

Woolford, K. & Dunn, S. (2013). Experimental archaeology and games: Challenges of inhabiting virtual heritage. *Journal on Computing and Cultural Heritage*, 6(4), 2–18.

Woolford, K. & Dunn, S. (2014). Micro mobilities and affordances of past places. Past mobilities: Archaeological approaches to movement and mobility. In J. Leary (ed.), *Past mobilities: Archaeological approaches to movement and mobility,* pp. *113–28.* Ashgate Publishing.

Zadora-Rio, E. (2003). The making of churchyards and parish territories in the early-medieval landscape of France and England in the 7th – 12th centuries: a reconsideration. *Medieval Archaeology*, 6097(March),1–19.

9 Conclusion

Conclusion: towards an understanding of deep mapping in the humanities

In 2008, Mary Spence, then President of the British Cartographic Society, argued in a lecture:

> Corporate cartographers are demolishing thousands of years of history, not to mention Britain's remarkable geography, at a stroke by not including them on [GPS] maps which millions of us now use every day. We're in danger of losing what makes maps so unique, giving us a feel for a place even if we've never been there.
>
> (quoted in Gray, 2008)

This statement may be true, but it highlights once again the risks of putting mapping technologies into silos based on a single type of use, and the importance of historical context. Spence's statement, made in 2008, was made at a time when GPS receivers were becoming widely used as navigational aids for drivers, but as maps for giving a 'feel for a place even if we've never been there', even then they hardly represented the *only* spatial shadows on the wall of Plato's Cave.

The view of the spatial humanities which emerges from the preceding eight chapters is of a field increasing in its self-awareness and critical reflection, and moving away from instrumentalist project-by-project assumptions about what technology can do. No longer is this only a domain where individual research problems are defined and solved in individual contexts by technologies that were not designed for the purpose, although many have defined, and continue to define, this process in terms of GIS, and of what GIS can and cannot do for particular subject areas. Rather, academics and practitioners are thinking creatively about the new research questions and opportunities GIS and the GeoWeb afford for the exploration of why place is significant in the human record. This concluding chapter seeks to draw out the most important factors which have emerged for the spatial humanities as digital methods of mediating and analysing place take hold further.

GIS is of course not the only technology to have formed a context for this process of methodological convergence. A key premise of this book is that both the GeoWeb and GIS have similar epistemologies, in that they force us to think about the discursive and fuzzy spaces and places which emerge from studying the human record in a particular way. It compels us to 'digitize place', rendering it into the binary structures of points, lines, polygons and raster data which a machine can compute, and an electronic network distribute. This represents both an abstraction and a fracturation of those spaces and places. Nonetheless, just as digitization and mark-up is a complex process, involving a chain of editorial, academic and critical decisions, which has led scholars of the textual Digital Humanities to think creatively about texts, so this process forces spatial humanists to think about place in different ways. It is in these different ways that the critical and intellectual value of the spatial humanities in the digital age lies.

The limitations of spatial technologies such as GIS that its increasing use in the academy and beyond has exposed in the way we treat qualitative human spatial information has led scholars in the humanities to invest considerable time, effort and resources into investigating both the theory and the practice of those technologies. The preceding chapters have offered an overview of how the spatial humanities have emerged, with special emphasis – following other recent reviews of the field – on digital mapping and GIS. However, they also argue that we cannot effectively critique these theories and practices without an understanding of the longer historical view. Technology, and the channels of communication which humans use to place-make, have *always* abstracted and fractured place in one way or another.

Every point of interaction in a social network at which place is described or represented provides an opportunity for that place to be rethought, re-understood and mediated in one way or another. The Internet and WWW have made this process faster, extending it across digital culture and, in the drive to make place as communicable and re-readable as any other sort of information made digital, it artificially homogenizes it by abstracting place into geometric and empirical binaries.

Probably most importantly, digital methods force us to (re)consider how we structure and represent place data in historical and humanistic sources. The informational and communicative structures which facilitate discourse in the humanities need to be enhanced and augmented, if they are to assist researchers to understand place in this new context. The concept of the gazetteer for example, and its pre-Internet history (a history which extends back to an era when the processes of producing and publishing such a document were an elite, or even colonial – activity), must be revisited. As most practitioners in the field recognize, the WWW gazetteer is first and foremost a convenient means for linking consistently expressed information, but this is simply an augmentation of the original purpose of the gazetteer. The major successes of current spatial humanities gazetteer projects have resulted from their ability to provide unexpected linkages between spatial datasets from different sources, and critical context to those linkages. They are *networks*, not isolated lists, and – like the Internet itself, as conceived by visionaries such as J. C. R. Licklider, Vinton Cern and Bob Khan – they use WWW standards to allow separate, independently functioning networks of information to communicate with one another seamlessly.

This segues into the next conclusion drawn here, which is that the spatial humanities must inevitably be seen as a product of the internet age, which was an 'information revolution', facilitating global communication across space as never before. The Internet was a chaotic, un-designed response to the sudden exponential growth in recorded information and the need to process it which, over the comparatively (relatively very) short time span of 50 years or so, shifted the focus of the information landscape from a set of nodes based around libraries, archives and collections (with their own specific physical locations, which framed and controlled access to them and their content) to a network of types of information which were disconnected from place. Vanevar Bush was not far wrong in his prediction that this information overload would resist the centralized and authoritative methods of librarianship applied in tightly bounded and controlled locations, which had been the norm hitherto; rather it would be organized algorithmically in cyberspace, in response to macro-level patterns in the ways in which millions of users interacted with the data. This is essentially what Google does today. And so it has proved with the GeoWeb, albeit on a smaller scale. The simultaneous and spontaneous actions of millions of users creating and contributing geodata in real time have often been portrayed as an open and democratic process but, as foreseen by thinkers such as Doreen Massey, it is acritical and unregulated in its nature, and this leaves it open to appropriation by big business and, possibly, unscrupulous governments and multinational actors. The lack of formal structures (such as gazetteers and libraries) does not set the neogeographies of 'Digiplace' free, rather it leaves them vulnerable to manipulation and the unconscious biases of their dominant demographics. One can argue that it is not only an intellectual, but also an ethical, requirement of the spatial humanist to be aware of such factors.

The problems of GIS-like analytical approaches in this world, based on reductive WWW-mediated interpretations of place when they step out of the comfortable and epistemologically safe zones of the expert-led project, are plain to see. That is not to say – and this is a very important point – that the application of GIS to a particular project, without reference to broader spectra of methods or approaches cannot be valuable. Any of the fields discussed in the preceding pages – archaeology, history, literature, textual criticism, performance – could point to individual examples of GIS or GeoWeb applications which have been responsible, directly or indirectly, for significant contributions to that field – but to understand what these technologies can bring to the *wider* study of the human record, and deeper understandings of human place in the digital world, then it is necessary to identify meaningful commonalities and ask what new methodologies emerge as a result. As noted in Chapter 1, the progression of academic applications in this area from a set of (largely) unconnected application-oriented approaches, to a broader 'methodological commons' is the main characteristic of the spatial humanities as a field, but digital approaches to place inside the academy cannot be completely disconnected from the roots they share with those outside.

Such engagement with bigger and wider ideas about human place in the physical and psychological worlds bring us back to the contributions of human geography and qualitative GIS, and from thence to deep mapping. A spatial humanities-wide approach to deep mapping, offers the possibility of transcending what Roberts calls the 'ballast of disciplinary persuasion' (Roberts, 2016: 2); which has been responsible for much of the fracturing of methodologies and approaches described above. It refers to the building of networks of different forms of spatial statement, observation or content. Deep mapping can be said to include the articulation of things such as fuzziness, observation, experience and emotion, and the intangible facets of engagement with the physical and corporeal world, which we encountered in Chapter 8. It means understanding how we express the processes of moving through a landscape at a particular time, in a particular way, and how that might deepen our understating of human interaction with place more broadly. It means communicating these things meaningfully as stories or arguments. This is very close to the definitions and high-level typology of spatial narrative discussed in Chapter 6. However, as Bodenhamer et al. have stated, a deep map is something that is *enacted*, not something that merely 'exists', as a conventional map might be described.

This means engaging creatively with the limitations of the spatial tools and methodologies with which humans handle place. Chapter 7 highlighted the fact that web mapping has disconnected the idea of place from place as it has been traditionally documented and structured, and this prompts us to ask further questions about why this separation is meaningful. For example, we talk instinctively about 'reading' maps, but the discussion in Chapter 8 in particular suggests that we do not 'read' place, it is something we engage with and experience via all our senses. Therefore do paper maps or 'conventional' computer maps transmitted through screen-based media, speak only to our sense of vision?

Yes and no. In his contribution to the 2015 volume, *Deep Maps and Spatial Narratives*, Harris debunks the notion that conventional types of map, such as the schematic representations described by Andrews, lack 'depth', saying: '[T]he term "shallow map" intentionally or otherwise implies a meaning of superficiality and inconsequentiality and that they are lightweight and lacking in substance. There is overwhelming evidence to disprove these … descriptions' (Harris, 2015: 30)

He urges to consider such maps as 'thin', rather than 'shallow', framing them (as in the discourses of critical cartography) as objects which are encultured and value-laden. We can agree with Harris that maps, be they schematic, VGI-based, top-down, WWW-based or plotted on paper, can be profoundly encultured objects, conveyors of ideas, of propaganda, of conscious or unconscious bias. But this does not free us from the obligation to see any such map as an agent of stasis. In other words, a 'thin' map, even if it is not 'shallow' in the sense Harris means, may well be the product of a discursive and culture-valued process (and, as described in Chapters 5 and 6, they can be authored either by an individual or a corporation). But when read by another party – the next point in the chain of the narrative's readership – it becomes, nonetheless, an inert view of that part of the world, perforce excluding any notion of dynamism or phenomenological engagement. The same goes for a description of a place in a text – and the rhetorical frameworks for understanding this are explored in Chapter 4. One might say the same for a performance, which becomes frozen as a memory. A *deep map*, however, is one which links other sensory experiences, and which transcends the limitations of individual media or analytical technologies by connecting what we learn with them to other things.

Place casts many shadows on the walls of Plato's Cave, and the historical overview given here shows that we receive them in myriad different ways. Looking to future, the philosophies of deep mapping give us a possible framework for understanding and, in some cases, formalizing them. Going back to Bodenhamer et al.'s definition:

> A deep map is a detailed, multimedia depiction of a place and all that exists within it. It is not strictly tangible; it also includes emotion and meaning. A deep map is both a process and a product – a creative space that is visual, open, multi-layered, and ever changing. Where traditional maps serve as statements, deep maps serve as conversations.
>
> (http://polis.iupui.edu/index.php/spatial-humanities/deep-maps-and-spatial-narratives, last accessed 26 July 2018).

Considering deep maps as being 'conversational', it is probably most useful to think of them as points at which spatial information or data is given and received. This can be points of communication between human beings, or between a human being and a place, a moment at which a description of place is read, or an image seen, and so on. It could be the point when, as described in the preface to his 'Cosmographie', Peter Heylyn was accosted by a 'tall, big gentleman' in Whitehall and informed that 'Geography is Better than Divinity' (Heylyn, 1677), another may be the point at which he commits his memory of the incident to writing; another might be the moment Heylyn's publisher set the text into type; another might be my reading that text 400 years later in the British Library. Given Heylyn's brief spatial description of the scene, we might approach my mental conception of the place with reference to the theories of rhetoric and spatial description presented in Chapter 4. Were I to blog or tweet about my experience of it, I would be remediating the 'place' of Whitehall into a format that the GeoWeb, as a medium for communication, can handle. And so on. A deep map can be understood as the trails of understanding, remediation, geometry and interpretation, that would, taken together, provide a 'deep' insight into Whitehall as a place.

As the response to Spence's argument above implies, the nature of geography on the WWW tempts us to overlook these trails of understanding, and to approach uncritically the fracturation that it perpetuates on the landscape, and the data which represent it. There is far more to human place than the points, lines and polygons of GIS and the GeoWeb; and this has implications for our understanding of the world. The foregoing passages highlight many of the tensions that this is creating in interpretation, culture, and our approaches to history and prehistory. Deep mapping online and offline, and the critical framework of the spatial narrative as an instance of a deep map combine different spatio-rhetorical intersections of understanding, that we must look for the meaning of dynamism and embodiment at the scale of structures, localities, and regions.

References

Gray, L. (2008). Map reading skills 'dying out due to internet and satnavs. *Daily Telegraph*. 28 August 2008. Available at: https://www.telegraph.co.uk/news/2639307/Map-reading-skills-dying-out-due-to-internet-and-satnavs.html

Harris, T. M. (2015). Deep geography – Deep mapping. Spatial storytelling and a sense of place. *Deep Maps and Spatial Narratives*, 28–53.

Heylyn, P. (1677). *Cosmographie in foure Books Contayning the CHOROGRAPHIE & HISTORIE of the whole WORLD and all the Principall Kingdomes Provinces, Seas and the Isles Thereof.* London: Anne Seile.

Roberts, L. (2016). Deep mapping and spatial anthropology. *Humanities*, 5(1), 5. Available at: https://doi.org/10.3390/h5010005

Index

For Product Safety Concerns and Information please contact our EU representative GPSR@taylorandfrancis.com
Taylor & Francis Verlag GmbH, Kaufingerstraße 24, 80331 München, Germany

www.ingramcontent.com/pod-product-compliance
Ingram Content Group UK Ltd.
Pitfield, Milton Keynes, MK11 3LW, UK
UKHW011017180926
475701UK00004B/794

* 9 7 8 0 3 6 7 6 6 1 5 0 2 *